A STING IN THE TALE

A STING IN THE TALE

My Adventures with Bumblebees

Dave Goulson

PICADOR NEW YORK

Printed in the United States of America. For information, address Picador, 175 Fifth Avenue, New York, N.Y. 10010.

www.picadorusa.com
www.twitter.com/picadorusa • www.facebook.com/picadorusa
picadorbookroom.tumblr.com

Picador® is a U.S. registered trademark and is used by St. Martin's Press under license from Pan Books Limited.

For book club information, please visit www.facebook.com/picadorbookclub or e-mail marketing@picadorusa.com.

Library of Congress Cataloging-in-Publication Data is available upon request.

ISBN 978-1-250-04837-0 (hardcover)
ISBN 978-1-250-04838-7 (e-book)

Picador books may be purchased for educational, business, or promotional use. For information on bulk purchases, please contact Macmillan Corporate and Premium Sales Department at 1-800-221-7945, extension 5442, or write specialmarkets@macmillan.com.

Originally published in Great Britain by Jonathan Cape, a division of Random House Group Limited

First U.S. Edition: May 2014

10 9 8 7 6 5 4 3 2 1

For Seth, my youngest son.
May there always be a flowery meadow
and the sight and sound of buzzing bees for him to enjoy.

Contents

Prologue

My interest in bumblebees and other insects dates back to the age of seven, when my family and I moved from a small semi-detached house on the edge of Birmingham's urban sprawl to a little village called Edgmond in Shropshire. My father had been brought up close by in the market town of Newport and, being a schoolteacher, he was keen that his two sons should get a good education. Newport had, and still has, a fine grammar school, the school which my father had attended and to which he hoped my brother and I would go, provided we could pass the entrance exam.

At seven I didn't much care about school, but I loved our new house. In hindsight it was rather ugly, with a rash of stone cladding and a hideous flat-roofed extension, but small boys don't worry about such things. The house was detached, and it had a much bigger garden than I was used to. There were large flower borders, apple and damson trees, a pond, two ancient wooden sheds full of cobwebs and vast spiders which gave me the willies, and enough room for my father to grow a fine patch of vegetables. Better still, the house was opposite open countryside. I had only to cross the village high street and jump a stone wall, and I was in a huge field, with a magnificent horse chestnut tree to climb. A grumpy dappled grey horse often stood in the tree's shade on hot summer's days, twitching its tail at the flies, and was prone to biting and kicking. In spring, the tree would teem with bumblebees visiting its pyramids of cream and pink flowers. The supply of bees meant that the flowers turned into plentiful conkers in late summer with which my friends and I would bomb passers-by while hidden in the dense green canopy of the treetop.

My father wasn't too interested in flowers; he allowed me to plant what I liked, so I put in lavender, buddleia and catmint to attract bumblebees and butterflies. I trained a honeysuckle up one of the old sheds to feed moths, and planted a male pussy willow to provide the bees with early spring food. I built a large rockery out of old bricks which I scavenged from a dilapidated farm building across the fields, carrying them home in a knapsack. I left spaces at the bottom for the bumblebees to nest in, and planted the top with bird's-foot trefoil to provide flowers for bees and tasty leaves for caterpillars of the common blue butterfly. I dug a bigger pond, and stocked it with newts, sticklebacks and all manner of other beasts from the local canal.

I have no idea where this all came from. My father was a history teacher who, to this day, can recite the chronology of English monarchs since William the Conqueror, with dates, and discern the age of a building from the shape of its windows or finials. Give him a bumblebee, however, and he hasn't a clue (although I have tried to educate him). My mother was a sports teacher, great with a rounders bat or a tennis racquet and fiercely competitive, but with no interest whatsoever in nature. She was not at all keen on creepy-crawlies of any description, and absolutely terrified of spiders. So I had to teach myself, using a range of identification books and natural history guides that my parents happily supplied me with; my father loved books of any sort.

The only adult whom I can recall actively encouraging my interests was a primary school teacher, the formidable Miss Scott. She was short and stocky, with thick brown curly hair and, having a short fuse, was prone to barking commands and reprimands.

My classmates and I were initially terrified – our previous teachers had been the sweet, gentle types that one imagines primary schoolteachers to be. But before long we realised that there was a merry glint in her eye, and that the stern front she presented was just that: a front. What is more, she loved to take us out looking for beasts and bugs; she showed us how to identify trees from the leaves, and how to place pitfall traps to catch beetles. She was particularly keen on pond-dipping. In my memory it seems that we went pond-dipping in the local canal every day (and it was always sunny). Our classroom soon filled with jam jars containing tadpoles, pond-skaters, ferocious dragonfly larvae, great diving beetles, millipedes, spiders and much else besides. The dragonfly larva was my particular favourite – this ugly, dumpy brown creature would lurk motionless at the bottom of the jar, waiting to be fed. Each day we would drop in a tadpole or a worm and watch, ghoulishly, as the dragonfly larva's face unfolded into telescopic pincers with which it snatched and devoured its unsuspecting prey.

By the following spring, my efforts to encourage wildlife in the garden were really beginning to pay off. I noticed huge queen bumblebees, fresh from hibernation, feeding on the pussy willow and lungwort. These bees had been asleep for seven months or so, since the previous July, so the spring feast I had grown for them was particularly welcome. Once satiated, the queens would fly low over the ground, searching for a suitable hole in which to nest. I noticed a white-tailed bumblebee queen investigating the space under one of the garden sheds, and she must have liked it, for weeks later her smaller workers appeared, flying out to gather food and coming back, half an hour later, with enormous balls of bright yellow pollen on their legs. I sat and watched them for hours, and noticed the nest traffic becoming busier and busier as the season went on and the number of workers rapidly grew. No bees ever showed any interest in nesting in my purpose-made bumblebee nest chambers beneath the rockery.

As summer approached, the garden began to swarm with wildlife. The buddleia was covered with small tortoiseshells, peacock butterflies, large and small whites, hoverflies and bumblebees. Pond-skaters and whirligig beetles fought territorial battles on the surface of my new pond, and an emperor dragonfly took up residence, perching on a tall purple loosestrife growing in the pond margin. It would zoom out to catch other flying insects to eat, snatching them mid-air with its bristly legs, and chase away any other dragonflies that tried to move in on its patch. I remain to this day amazed at how quickly wildlife appears in a garden if given just a little encouragement.

On one occasion, after a heavy summer rainstorm, I found a number of bedraggled bumblebees clinging to my buddleia, and decided to dry them out. Unfortunately for the bees I was, perhaps, a bit too young to have a good grasp of the practicalities. With hindsight, finding my mum's hairdryer and giving them a gentle blow-dry might have been the most sensible option. Instead, I laid the torpid bees on the hotplate of the electric cooker, covered them in a layer of tissue paper, and turned the hotplate on to low. Being young I got bored of waiting for them to warm up and wandered off to feed my vicious little gerbils. Sadly, my attention did not return to the bees until I noticed the smoke. The tissue paper had caught fire and the poor bees had been frazzled. I felt terrible. My first foray into bumblebee conservation was a catastrophic disaster. This did not bode well for the future – but at least I had learned that there is an upper temperature limit beyond which bumblebees are not happy. As we shall see later, a similar principle explains why there are few bumblebees in Spain.

I was an avid fan of Gerald Durrell's books, particularly those about his childhood growing up in Corfu, collecting all sorts of exciting animals and keeping them in his bedroom. He had owls, snakes and turtles – and, what is more, he never had to go to school (he was taught at home by an eccentric tutor who was more interested in swordfighting than algebra). He even had a donkey to carry all his collecting nets and jam jars. Deeply envious, I did my best to follow in his footsteps, making do with the slightly more mundane fauna of Shropshire. I badgered my poor parents into letting me keep an array of pets, starting with guinea pigs, rabbits, hamsters and mice. My brother and I mercilessly wore our parents down until they agreed to let us have a dog, a lovely black Labrador-cross puppy that, with a total failure of imagination, we named Spot after the white spot on her back. As she grew, this spot rapidly disappeared, which made her name the cause of occasional confusion. Nonetheless, she was an incredibly soft and tolerant dog who put up with our endless teasing and was a great companion in our romps in the countryside.

After the novelty of my traditional pets wore off, I moved on to more exotic tropical fish, leopard frogs, red-eared terrapins, garter snakes and anolis lizards. I had my own bedroom with a view of the chestnut tree, and I filled this room with home-made boxes and tanks from which all but the most dim-witted creatures invariably escaped. My garter snakes spent more time out of their tank than in it. In desperation I tried using sticky tape to hold the lid down, with unfortunate consequences. One of the snakes still managed to push up the lid, but then became stuck to the tape and in its attempts to disentangle itself became hopelessly wrapped up in a ball of tape; it took me hours to tease it apart. I resigned myself to regular hunts for escapees, and it is quite possible that a garter snake is, to this day, living somewhere under the floorboards of that house.

For one birthday, I was given a small aviary for the garden which I stocked with budgerigars and a pair of beautiful Chinese painted quails. As an adult I find keeping birds in cages cruel (especially large parrots in small cages indoors) but as a boy I wasn't worried by such sensibilities. I loved sitting in the aviary with the birds flying about my head. Before long the budgies started to breed, and I was able to supplement my pocket money by selling the surplus stock (the quails also laid plenty of eggs but they never seemed to hatch). Baby budgies are spectacularly ugly bald creatures with oversized heads. Normally they rapidly grow feathers and become rather more cute, but one poor chick seemed unable to, and as it grew remained almost entirely bald. Eventually it attempted to fledge from the nest and leapt out, falling like a stone to the floor. Undeterred, it clambered back up the netting using its beak and feet and joined the other budgies on the highest perch. Every now and then the poor little mite would gamely hurl itself into the air, flapping its tiny pink arms, and thud once again to the floor. It lived for six months or so but stood little chance when winter arrived.

My charges had a worryingly high mortality rate. One Sunday morning, my mother was in the kitchen rustling up one of her legendary pies (she is an excellent but very traditional cook, always serving up meaty dishes with potatoes and vegetables, followed by a hefty hot pudding such as a fruit crumble or spotted dick with custard). I must have been at a loose end and getting in her way, so she pointed out that the fish tank in my bedroom was in dire need of a clean – the glass had become green with algae, so that the fish were barely visible. A little while later I was dutifully scrubbing the glass inside the tank, my arm immersed in the warm water, when my mother called up, 'Dave . . . What's burning? You aren't lighting matches again are you?' Before starting to scrub, I had lifted out the electric heater, encased in its glass waterproof tube, and laid it on a wooden cupboard to one side. It hadn't occurred to me to unplug it, and not being in water it had become hot and was burning into the top of the cupboard. (I never fathomed how my mother was able to smell burning so quickly and from such a long way away.) Without thinking, I lifted the heater by its cable and tossed it into the tank. Of course, very hot glass and cool water are not an ideal combination, and the heater tube shattered with a bang, exposing the electrical element to the water and electrocuting all of my fish. They quivered and spasmed in the water (thankfully I didn't shove my hand in the tank to pull the heater back out), and by the time I had pulled the plug from the socket they were all very much dead.

There were many other such disasters. Perhaps the most traumatic involved my quails. These lovely little creatures scurried around on the floor of the aviary scratching for food. The male had beautiful black and white markings on his face, and a rather splendid plume on top of his head. The female was more drab but delicately marked with dark speckles. They were inseparable, behaving as if glued together side by side, and often grooming one another. I preferred them to the budgies, which I had eventually concluded were decidedly raucous, uncouth and gaudy beasts (perhaps my view was coloured by the savage pecking they gave me whenever I had to handle them). Now, Shropshire is a cold county in winter, as my bald budgie discovered. It is a long way from the warming influence of the sea, and often records the coldest night-time temperatures in England. After one particularly cold night I went out in the early morning to feed the birds in the aviary, and was surprised to see the budgies attacking the quails. Both quails were struggling on the snowy ground, each with two or three budgies perched on them and tearing mercilessly at their feathers with their jaggedly pointed beaks. I rushed in and shooed the budgies away. The poor quails seemed unable to stand, but were very much alive. I picked them up, one in each hand, and took them indoors. On the kitchen floor, it became obvious what their problem was. Whenever they tried to stand, they simply toppled over. Closer inspection revealed that they had no toes; they had both suffered from frostbite in the night, and their toes had simply dropped off. Their legs now ended in stumps, no use at all for standing up or walking. Distraught, I did not know what to do. In a flash of desperation, I tried to fashion them prosthetic feet from plasticine and matchsticks, but this was not a triumph so I laid the birds, still struggling to come to terms with their new prostheses, in a cardboard box with some food and went to school.

When I came home, the situation had not improved. The birds had not miraculously grown back their toes, or worked out how to use their new plasticine-and-matchstick feet. They were just lying there, looking a little more feeble. The harsh reality dawned: my quails were not going to get better. They could not be fixed. I had felt terribly guilty that my bald budgie had probably frozen to death, and it was clear that it would have been kinder to have given it a swift death. With this in mind, I decided that there was only one thing for it.

I cannot remember why I decided against enlisting the help of my parents to take the poor birds to the vet at this point. A quick lethal injection would have been the most sensible solution, but small boys are not logical. Instead, I got my dad's axe from the shed. It was a full-sized, grown-up's axe, way too big for me at the time. I took the birds to the bottom of the garden, and laid them next to each other on the grass. I figured that it would be best to deal with them both at once, rather than dealing with one while the other looked on. They lay there, looking up at me, their eyes still bright, their stumps kicking ineffectually. I hefted the axe on to my back and took a huge swing. The head of the axe buried itself in the lawn, just in front of the beaks of the startled but otherwise unharmed birds. I had been aiming to sever both their heads in one blow. I eventually managed to pull the axe out of the ground, and tried again. Success! More or less. I didn't so much sever the heads as chop both birds clean in half, but the end result was much the same. I dug a small hole next to my rockery and laid them to rest, roughly reassembled and side by side, as they had spent their lives.

I could go on. I could mention the awful fate of my axolotl, or my botched attempt to perform corrective surgery on a badly injured rook. Suffice it to say that being one of my pets was a dangerous business.

As well as amassing a diverse array of living creatures, I became an avid collector. I am embarrassed to admit that this started with birds' eggs. In the 1970s in rural England this was a very common hobby for boys. Many of my friends collected eggs, and we would vie with one another to obtain unusual specimens. My father showed me how to blow the eggs; he had collected them himself as a boy, along the same hedgerows that I now searched. One grinds a tiny hole at each end by spinning a pin between one's fingers while pushing the tip against the shell. The idea is to then blow on one end, forcing the contents out through the opposite pinhole. Easy enough with a chicken's egg, but incredibly fiddly with the tiny white-and-brown-speckled egg of a wren. My prize specimen came from a mute swan. When out 'egging' along the local canal bank with my friends Les and Mark (or 'Butt' as we knew him, for reasons long forgotten), we spotted the egg lying in an abandoned nest in a reed bed near the opposite bank. The rest of them had long since hatched and the parents and cygnets were nowhere to be seen. Without hesitation we threw our jumpers and T-shirts off, knowing that the first to get there would win the prize. Butt and Les started peeling off their jeans, but I just leapt in half-clothed and beat them to it. The egg was putrid inside; when I pushed the pin into it a stream of creamy, lumpy goo erupted from the end, squirting into my face and smelling to high heaven. Blowing out the rest of the contents was a memorable ordeal, which my long-suffering father helped me with in the end as I had turned green from the smell. The egg was eventually placed, still rather whiffy, in pride of place in the centre of my display case on my bedroom wall.

Modern readers will be horrified by all of this. Egg collectors are now only one small step above serial killers in the social hierarchy (in fact, I suppose in a sense they *are* serial killers, so fair enough). It is true that most of the eggs I collected were alive when I took them, unlike the swan's egg. I do not defend egg collecting; I certainly would not allow my three boys to do it. But I did learn an awful lot about natural history by spending my days hunting for eggs. We only ever took one from a nest, and did our best to disturb it as little as possible. This does not, of course, make it right. Collecting the eggs of extremely rare birds is clearly a heinous crime, and I am glad that I never managed to find anything particularly rare. But I sometimes think that we are poor at keeping perspective on our activities, and those of others. How many condemn egg collecting, for instance, while allowing their pet cat to roam unfettered? (Domestic cats kill millions of birds and small mammals each year.)

From eggs I moved on to collecting insects, starting with butterflies. My mother, bless her, was not keen on this – but I persuaded her that I would only take a male and female of each species, and could not do too much harm. To start my collection I bought a dead, dry but very beautiful tropical swallowtail from a butterfly farm in Dorset called Worldwide Butterflies. It arrived in a paper envelope inside a small cardboard box which I opened with great excitement. What I hadn't anticipated was that the specimen would not have been 'set', which is to say that its wings were folded shut, and it did not have a pin through it. I tried to open the wings, not understanding that this is impossible with a dry butterfly; they are incredibly brittle and delicate. The wings snapped off along with most of the legs as I clumsily tried to arrange it in an attractive position. I was left with a very sad collection of body parts. Disheartened, I managed shortly afterwards to get hold of a second-hand book, *Studying Insects* by E. B. Ford, which explained where I had gone wrong. To pin and set a butterfly with the wings flat and beautifully symmetrical, as they are always displayed in museums, it must be freshly killed, or if it is dry it must first be 'relaxed' by putting it in a tin with moist tissue paper for a couple of days (no longer or it goes mouldy). When soft and damp the butterfly can be carefully pinned and arranged in whatever position is desired. Once it dries, it will remain fixed in position for ever, so long as it does not get damp again.

Studying Insects also explained how to make a killing jar by filling the bottom of a large jam jar with crushed laurel leaves; when crushed, the leaves release cyanide, which smells strongly and sweetly of marzipan (even knowing it was poisonous, I couldn't resist having a good sniff every now and again). A few minutes inside the jar is enough to send a butterfly into a permanent sleep.

I also tried constructing a butterfly net from a wire coat hanger and a pair of my mum's stockings, but this was hopeless, and without a net it was almost impossible to catch anything. Eventually, I discovered the address of a company named Watkins & Doncaster, based in Hawkhurst in Kent. They billed themselves as 'suppliers of entomological equipment'. I wrote to them, and a few days later received their catalogue through the post.

This was a seminal moment in my life, a turning point from which I have never looked back.

I had just arrived home from a game of mini-rugby, so I guess I was eight years old. I was covered in mud, so I took the catalogue up to read in the bath. The Watkins & Doncaster catalogue was the most wonderful thing I had ever seen. Fat, it contained page after page of illustrations of the most amazing paraphernalia: insect nets, pond-dipping nets, pillboxes, cages, tubes, magnifying glasses, malaise traps, microscopes, setting boards, moth traps, pooters, beautiful mahogany insect cabinets. At the end was a section on taxidermy, which contained such entrancing objects as a brain scoop, bone cutters, and a vast selection of glass eyes. I was transfixed, amazed. This was a whole new world. Moreover, there were obviously lots of other people out there like me! I wanted to buy more or less everything in the catalogue, but my pocket money placed severe limits on what I could afford. Nonetheless, my first purchase was a full-sized, professional kite net which cost me £16, a fortune to an eight-year-old boy, and I was immensely proud of it. It was nearly as tall as me, with a stout brass handle, a rigid metal frame and a soft and very deep black net. With this, I felt I could catch almost anything.

My butterfly collection slowly grew, as did my collection of books about butterflies and other insects. My first catch was a terribly tatty painted lady, her wings torn from the long migration from Morocco. I soon added a meadow brown, large and small whites, a gatekeeper, speckled wood, small tortoiseshell, red admirals, common blue and peacock. The beauty of these creatures takes my breath away to this day; I still have the specimens, in the top drawer of an insect cabinet which I was only able to afford three decades later. I also learned to search for the eggs and caterpillars, which meant finding out what the caterpillars ate, and also how to identify the plants. With a little care it is easy to rear caterpillars into adult butterflies; that way one gets beautiful fresh specimens to add to one's collection, and the surplus can be released. I picked up an enormous amount of knowledge.

From butterflies I expanded my interests to include moths. Most moths fly at night, and to catch them there are two popular approaches. One is to go 'sugaring'. This involves boiling up a fantastic brew of black treacle, beer, brown sugar, vanilla essence, pear drops, rum or brandy, and pretty much anything else one fancies so long as it adds to the heady aroma. Every moth collector has his own highly secret recipe, or so it seems. Whatever the mixture, the end result should be a thick gloopy liquid that smells so strong that it makes one's eyes water at fifty paces. This is then painted at dusk on to fence posts or tree trunks. The idea is that moths find the smell irresistible and are drawn to land and drink the sugary syrup; they become hopelessly intoxicated by the alcohol, and then sit there in a stupor ready to be snatched up by the eager moth collector. I stank out the house brewing up various versions of this, and got through much of my mum's sugar, treacle and food flavourings and a lot of my dad's alcohol. The end results were disappointing. Earwigs appeared to be the only creatures that were consistently attracted; I sometimes had hundreds of them swarming over my sugar patches, getting stuck in the goo as they climbed over each other in their feeding frenzy. Hardly a single moth appeared. I also found it slightly nerve-racking wandering around the local fields at night on my own (not least because my father regularly let me and my brother stay up on Saturday nights to watch Hammer House of Horror movies, and my overactive imagination conjured up a vampire in every shadow). On one occasion I was checking the sugar patch on a large ash tree when a tawny owl decided to screech just above me. Although I knew it was an owl, I had great difficulty resisting the temptation to sprint straight back home, and my heart didn't stop hammering in my chest for a good ten minutes.

There is an alternative and more convenient way to attract moths: a light trap. *Studying Insects* explained the principle: moths are attracted to candles and any other light sources. Hence this sort of moth trap involves a bright light hung above a container a bit like a lobster pot. The moths are drawn to the light, blunder into it and fall down through a funnel into a large dark container usually stuffed with egg cartons, which they seem to like to sit on. This sounded much easier and less scary than traipsing round the fields in the dark with a bucket of treacle, so I decided to give it a go.

I rigged up a 100-watt light bulb over a home-made cardboard funnel, itself sitting on a plastic bucket, turned it on before going to bed and eagerly awaited the morning. I dashed down at first light to survey my catch. Disappointment: nothing but a couple of wasps and a tiny brown 'micro' moth as I now know they are called. I tried for a couple of weeks, but with little success. After some research, I gathered that ultraviolet light was best for attracting moths. By chance my mother had a rather odd and old-fashioned heat lamp used to treat muscular injuries, something she had possessed ever since she was in college training to be a sports teacher. It resembled an enormous Anglepoise lamp, but with two very fancy-looking bulbs, one of which produced infrared heat, and the other ultraviolet light. To this day I've no idea why anyone thought it was a good idea to give injured body parts a jolly good tan as well as a blast of heat; presumably skin cancer was not well understood at the time. Anyway, I'd never seen my mother using it (probably a good thing) and I figured she wouldn't mind if it was cannibalised in the name of scientific research. The only problem was that it wasn't possible to turn on the ultraviolet lamp without the heater element. Undeterred, I rigged up both bulbs next to one another above my home-made bucket trap, and left it on for the night. The next morning, I came down to a qualified success. The UV lamp had attracted a lot of moths, but unfortunately they had been frazzled to a crisp by the heat lamp: my trap was full of charred moth bodies. Not quite what I was after. In frustration, I attempted to rewire the lamps to separate the two bulbs. I don't think I had started physics at school by that age (I was about nine years old), so this was inevitably something of a long shot. When I flicked the modified light on, this time with only the UV bulb connected to the power, there was a loud bang. The UV bulb shattered. I reassembled Mother's lamp and put it back in the cupboard, hoping that she would never notice. Of course she did. It was many years before I saved up enough money to buy a proper 'Robinson's mercury vapour moth trap' (an absolutely marvellous device, by the way, which lights up the entire neighbourhood with an eerie glow and attracts moths from miles away). In the meantime, my moth collection grew rather slowly.

I was not fully aware of it at the time, but my childhood coincided with a catastrophic period in the history of the British countryside, at least from the point of view of a butterfly or bumblebee. Shropshire may sound idyllic, but this is misleading. It was and is a relatively rural, green and pleasant part of Britain, but it is not the haven for wildlife that it might once have been. I moved there in 1972, and left for university in 1984. At weekends I would often walk with my friends to the Shropshire Union Canal about two miles away across the countryside, searching the hedges for birds' nests along the way. When I started, this walk involved crossing fifteen fields, each bounded by a hedge. By the time I left for university, the walk involved crossing one field – a huge one. The hedges in which I used to search for birds' eggs had been ripped out, one by one. A large part of the canal itself had been filled in, covered in topsoil, and was now just a part of the arable expanse. Where once a bumblebee would have been able to find brambles in the hedges, cowslips in the hedge banks and marsh woundwort on the sides of the canal, there was now only a sea of cereals, a monoculture stretching across the landscape. These changes occurred almost everywhere in lowland Britain, sweeping across Western Europe.

These changes drove the decline and, in some cases, the extinction of many creatures, and our countryside is a much poorer place because of it. But the battle is not lost. We have slowly, tentatively, started to find ways to undo the damage. Scientific studies are revealing how best to combine efficient farming with looking after the countryside. A range of payments is available to farmers to support them in encouraging wildlife. The British have a peculiar and unique love of the countryside and the animals and plants which inhabit it, and there is a huge groundswell of support for conservation. To tap into this, in 2006 I launched the Bumblebee Conservation Trust, a charity devoted to saving our bumblebees, and to my delight the Trust has flourished. It now has over 8,000 paid-up members, and is creating flower-rich habitat for bumblebees across Britain from Kent to Pembrokeshire to Caithness. Most of our wildlife clings on, and with our help it can recover. Sometimes even species which have been lost entirely might one day return. But that is the subject of Chapter 1 . . .

CHAPTER ONE

The Short-haired Bumblebee

In the 1870s, New Zealand farmers found that the red clover which they had imported from Britain, as a fodder crop for horses and cattle, did not set much seed. As a result, they found themselves having to continually import more seed from Europe at considerable expense, rather than collecting and sowing their own. In the end a solicitor named R. W. Fereday worked out the cause of the problem. Fereday had emigrated to New Zealand in 1869 and, aside from his legal work, was a keen entomologist with a particular interest in small moths. It was Fereday who realised, while staying on his brother's farm, that the problem lay in the absence of the bumblebees which normally pollinated the clover back in Britain. The problem was taken up by Frank Buckland, Her Majesty's Inspector of Fisheries at the time, whose remit seems to have extended well beyond fish. He wrote back to England with a request for bumblebees to be sent on the steamships which regularly plied between Britain and New Zealand. The first, rather ill-thought out, attempt to do so involved a Dr Featherston digging up two carder bumblebee nests in late summer and sending them to the Honourable John Hall of Plymouth, New Zealand, in 1875. They arrived in January and, inevitably, were all dead. Bumblebee nests naturally die out in September, and in any case there were no flowers on the ship for them to feed on, so this scheme was doomed from the start.

Eight years later the idea was revived with rather more competence. A Mr S. G. Farr, secretary of the Canterbury Acclimatisation Society (of whom more later), contacted Thomas Nottidge, a banker from Maidstone in Kent, asking for more bumblebees to be sent. (They also asked him for a few hedgehogs while he was at it – as you do.) So it was that, in the autumn of 1884, Nottidge offered a bounty to farm labourers for every hibernating bumblebee queen that they could find. Hand digging, clearing and widening of ditches was a common autumn and winter practice on arable farms when there wasn't much else to keep farm labourers busy, and these labourers often turned up the plump hibernating queens as they dug, suggesting that queen bees particularly like to hibernate in ditch banks. As a result, a total of 282 queens were obtained and placed on the SS *Tongariro*, one of the first steamships to be built with a refrigeration unit. This was essential as the hibernating queens would otherwise have become too warm when crossing the equator, and would have woken up and quickly died. The *Tongariro* left London in December 1884 and arrived in Christchurch on 8 January 1885 (high summer in New Zealand). When they were warmed up, forty-eight queens proved to still be alive. They were fed with honey and flew away. A further consignment of 260 queens was sent that same January on a sister ship, the SS *Aorangi*, and arrived on 5 February. Of these, forty-nine were still alive and were released.

We have no idea what species of bumblebee these ninety-seven queens belonged to, or how many survived long enough to build a nest and produce offspring. What we do know is that some thrived in their new home for, by the summer of 1886, bumblebees were seen up to 100 miles south of Christchurch. Indeed, by 1892 bumblebees had become so common in some areas that honeybee keepers feared they might become a pest.

British bumblebees flourish in New Zealand to this day. On

their long boat trip they also left behind many of the diseases and parasites that attack them in their native land, which probably helped considerably. The species that survived are an odd selection. We might have expected them to be the most common Kent species, but either our most common species were not included or they failed to survive. The four now found in New Zealand are the buff-tailed bumblebee, the garden bumblebee, the ruderal bumblebee and the short-haired bumblebee. Of these, the buff-tailed is by far the most common – they are everywhere, from the gardens and parks of Christchurch to the spectacular fjords of Milford Sound, where I have seen them feeding on the flowers of the gigantic New Zealand flax. The short-haired bumblebee is the least common, but if you know where to look, they can still be found in central South Island.

Sadly, two of these species have not fared so well in the UK. The ruderal bumblebee was once known as the 'large garden bumblebee' because it was a familiar sight in gardens throughout much of England. Nowadays the ruderal bumblebee is an exceedingly rare creature, found only in a few places in the East Midlands and East Anglia. The short-haired bumblebee has fared even worse. One hundred years ago they were common in the south and east of England, but during the second half of the twentieth century their numbers plummeted. By the 1980s they were known only in a handful of places, and one by one, those populations disappeared. The last individual was caught near Dungeness in 1988; it fell into a pitfall trap used to monitor beetles and drowned. No one has seen any since.

Of course you will have worked out why these bees disappeared. It happened while I was growing up. When I was born in 1965 the short-haired bumblebee was still quite widespread, although not as far north and west as Shropshire. By the time I went to university in 1984 it was nearly extinct. I never saw one before they vanished.

Here's why: it's Adolf Hitler's fault. To be absolutely fair, it wasn't entirely his fault, but he has to carry some of the blame. One hundred years ago, farming was not mechanised. Without mechanisation, fields tended to be small. Farmers depended on horses for power, and horses love to eat clover, so most farmers grew clover. Bees also love clover. Both the horses and other farm livestock needed hay for the winter, so most farmers had hay meadows. These were permanent features of the farm, cut once or twice a year, and sometimes grazed a little in the milder winter months. Artificial fertilisers weren't available, so apart from a bit of animal dung the meadows were not fertilised. In the low-nutrient soils of hay meadows, wild flowers flourished, particularly those with symbiotic root bacteria that could trap nitrogen from the air and so didn't need nutrient-rich soil. The main family that can do this is that of the legumes: vetches, trefoils and clovers (and also our garden peas and beans). Bees love them all.

Arable crops need fertile soils. The traditional way to maintain soil fertility was to grow crops in rotation. For many centuries, European farmers used a three-year rotation of rye or wheat followed by oats or barley, then letting the field lie fallow in the third year. In the eighteenth century, a British agriculturalist named Charles Townshend promoted a four-year rotation, using wheat, turnips, barley and clover in succession. The nitrogen fixed by the clover boosted soil fertility in the following years, increasing yields, and the scheme was widely adopted. So, imagine Britain a hundred years ago; a patchwork of small fields, cereals and root crops intermixed with clover leys and permanent hay meadows. No artificial fertilisers, no pesticides. Lots and lots of happy bees.

Then roll forwards a few years. The internal combustion engine had by now provided farmers with an alternative to horses, in the form of tractors. The booming motor industry demanded oil, and the petrochemical industry that grew up on its back made it possible to synthesise cheap nitrogen-based fertilisers. These greatly boosted crop yields and removed the need for rotations, so clover leys were abandoned. Moreover, horses were no longer needed, so no clover was necessary for feeding them.

Silage making is an alternative approach to providing winter fodder for livestock. Where hay requires a dry period for harvesting, meaning that wet summers can be a disaster for farmers dependent on it to feed their animals, the grass for silage can be cut even when it is wet. With the addition of cheap fertilisers to hay meadows, the grass grows much more quickly and so can be cut for silage many times during the spring and summer, providing a larger and more reliable supply of winter fodder. An unfortunate side effect is that adding fertilisers to hay meadows quickly results in the disappearance of most of the wild flowers. The clovers and other legumes, which used to gain an edge from their ability to fix nitrogen from the air, lose their advantage when nitrates are poured on to the ground, and cannot compete with fast-growing grasses.

None of this sounds good for bees, for fewer clover leys and fewer hay meadows means fewer flowers. So where does Hitler come in? By the advent of the Second World War, farming in the UK was changing, but only slowly. The techniques for growing more food were available – tractors, fertilisers, silage – but farmers tend to be traditionalists at heart and often farm as their parents farmed. There was no great pressure to change. Then, in 1940, Britain found itself isolated. No food could be brought over from mainland Europe. Obtaining supplies from across the Atlantic was perilous, with U-boats taking a heavy toll on shipping convoys. Before the war, Britain had been importing about 55 million tons of food each year. Suddenly, being able to supply enough food for our substantial population living on our small and crowded island became terribly important. As a result, the government launched a 'Dig for Victory' campaign, encouraging everybody to dig up their lawn and grow as much food as possible. At the same time, farmers were encouraged to use every measure available to maximise food production. Patches of land which had previously been deemed too small to bother with were now ploughed and sown with crops, hedges were ripped out, marshes were drained. Between 1939 and 1945 the area of land used for food production rose by 80 per cent.

From a bumblebee's perspective, the war era led to some other unfortunate developments. The chemical dichlorodiphenyltrichloro-ethane (usually known as DDT) was first made in 1874, but its incredibly high toxicity to insects wasn't discovered until 1939, when the Allies were desperately searching for chemicals to help combat the mosquitoes that spread malaria and typhus among the troops fighting in Asia. By 1945, DDT was readily and very cheaply available as an agricultural insecticide. It was twenty years before its long persistence and devastating effects on the environment began to be recognised. Also during the war, research in Germany into chemical warfare agents (nerve gases) led to the development of a range of organophosphate chemicals which were also highly toxic to insects. These too became available to farmers shortly after the war, providing them with a growing armoury of pretty unpleasant compounds with which to combat insect pests.

After the war ended, the policies which had been introduced to increase food production continued. Food rationing ended in 1954, but farmers carried on receiving financial incentives to increase production until the 1990s. Over a period of fifty years, we therefore destroyed almost all the flower-rich habitats in the UK, and 98 per cent of our lowland hay meadows disappeared. The short-haired bumblebee died out because the habitats in which it lived were swept away. It wasn't all that fussy, it just needed enough flowers to feed on. No flowers equals no bees. It is not rocket science.

Luckily for the short-haired bumblebee, Hitler didn't have the same impact on New Zealand. In fact there is a certain irony that this species now survives in the clover rich pastures that man has created in New Zealand by clearing dense native forests which would have been entirely unsuitable for bumblebees, whilst back in its native land we have been busy destroying its habitat. While the short-haired bumblebee has been away, many changes have taken place in Britain. Yet by the 1980s and 1990s it was becoming all too obvious that most of our wildlife was in rapid decline, and that in the long term what we were doing to the countryside might not be sustainable. Farms need flowers to support the bees that pollinate our crops, and they need predatory beetles, wasps and flies to eat the pests that eat the crops. So it was that schemes were introduced to pay farmers for encouraging wildlife on their land. Farmers can now get funding to re-sow the wild flower meadows and replant the hedges that only thirty years ago they were paid to remove. It might just be that we have turned a corner. But if British wildlife is very slowly beginning to recover, it can certainly do with a helping hand.

The presence of British short-haired bumblebees in New Zealand provided a unique and exciting opportunity to give our beleaguered wildlife a boost, and to act as a flagship for conservation efforts for bees and flowers. Why not bring them back from New Zealand? Could we once again have short-haired bumblebees buzzing across the British landscape?

One obvious obstacle is that we didn't know much about this creature. There was very little in the way of studies of short-haired bumblebees before they went extinct in the UK. There would be no point in bringing them back and then watching them die out again for exactly the reason they died out in the first place. We would need to be certain that there were now enough of the right flowers for them to feed on, but we had scant records as to the flowers they favour.

So it was that in January 2003, I found myself in New Zealand with a friend and colleague, Mick Hanley, in search of the short-haired bumblebee. Mick is a stocky, ginger-haired beer-drinking Black Country lad, who did his PhD on slugs (he prefers to call it 'seedling herbivory', but a lot of slugs were involved). At the time he was working for me on an ill-fated project to find a means of controlling fly outbreaks on landfill sites, but he is an excellent botanist and shares my enthusiasm for pies, so he made a great travelling companion. Our mission was to find out more about the food plants and habitats of the elusive short-haired bumblebee, to pave the way for an attempt at reintroduction. We needed to know which flowers it favoured for collecting pollen, which for nectar, and what habitats it was found in. Ideally, we wanted to find out where it liked to nest. Once we knew these things, it might be possible to recreate suitable habitat in Britain. Good reasons though these were, the prospect of escaping the northern winter for New Zealand summer sunshine was also attractive.

We set out from Christchurch in a tiny and rather flimsy hire car, heading south-west towards the centre of South Island which was where, we were told, the short-haired bumblebee had its hideaway. New Zealand is a land of marked contrasts. Christchurch sits on the Canterbury Plain, a rather monotonous and absolutely flat stretch of farmland covered in a neat grid of rectangular fields and a scattering of small, pretty but unremarkable towns. As we hurtled along the dead-straight road – Mick has a habit of driving ludicrously fast – ahead and to the right we could see in the distance the snow-capped peaks of Mount Cook National Park. Every few miles we crossed rivers full of snow-melt flowing down from the mountains to the sea, their shingle banks clothed in yellow tree lupins. We stopped for the night in the pleasant market town of Geraldine, and the next day, with me at the wheel, we proceeded at a slightly more leisurely pace along increasingly windy roads as our route started to climb into the foothills of the Mount Cook range. There, the neat arable fields gave way to sprawling sheep ranches and scree-covered hillsides glowing purple with viper's bugloss. According to the old records, we were entering short-haired bumblebee territory. Every few miles we stopped and searched, finding that buff-tailed bumblebees were common everywhere, and ruderal bumblebees almost as abundant. The latter was a real treat as I had only ever seen one small worker before, on Salisbury Plain. At this time of year in New Zealand the ruderal queens were still on the wing; they are absolutely huge, the biggest British species, more like flying mice than bumblebees.* They are also unusual in that they come in many colours (most bumblebee species are fairly uniform). Some are entirely jet black, others have a range of brown or yellow stripes and white or brownish bottoms. It would be wonderful if they could one day become a common sight in British gardens as they once were.

In part because they are so variable in colour, ruderal bumblebees are quite hard to separate from garden bumblebees – the two species are very closely related. Both are not dissimilar to short-haired bumblebees. As Mick and I spent several hours catching bees and staring at them in an attempt to decide what they were, I couldn't help thinking that the folk who introduced bumblebees to New Zealand could have thought to make life easier for future entomologists by introducing only readily distinguishable species. The accepted technique is to place the bee in a clear plastic tube and then push a ball of tissue paper in so that the bee becomes

* One day I must go to Chile to see the legendary *Bombus dahlbomii*, the world's biggest bumblebee, a monstrous fluffy ginger beast that lives in the high Andes and on the chilly tundra of Tierra del Fuego.

trapped and cannot move. They don't much like it but it doesn't do them any harm, and you can then get a good look at them. We usually used urine sample tubes – they are cheap and do the job so I always carry a few with me in summer, but it is sometimes a bit embarrassing when a handful of them tumble out of your pocket in a public place, suggesting that you have a serious urological problem. After a lot of staring at slightly squashed bees we decided that we could distinguish between garden and ruderal bumblebees (it is all to do with the shape of the second yellow band on the back of the thorax, if you really want to know). None of them were short-haired bumblebees. According to the books, the latter are supposed to have faint greenish-brown stripes on their abdomens, but no matter how hard we stared we couldn't find anything that matched this description.

On the third day we arrived at the stunningly scenic Lake Tekapo. The lake is filled from glacial meltwater which, due to the tiny fragments of crushed rock suspended within it, is the most striking icy blue. The lake stretches for 20 miles to the foot of Mount Cook whose snow-capped peak reflects beautifully in the chilly waters. The shingle-covered shores of the lake and the surrounding meadows were awash with so many wild flowers – towering mauve spikes of viper's bugloss, dense lush stands of purple and yellow lupins, and luxuriant tussocks of red clover – that this was surely bee heaven. We jumped from the car and excitedly explored the flower patches, catching bees by the dozen. Still no short-haired bumblebees. After a couple of hours of searching I saw a large bee that looked a bit different, feeding on the bugloss. She was big, a queen, but a little smaller than the ruderal queens. She wouldn't keep still, but I was sure that she had some greenish-brown stripes.

Now, I pride myself on my proficiency with a butterfly net. I have been using one for years, and I don't often miss. There is a knack to stalking an insect, then timing the sweep and flicking the end of the net over so that one's prey is trapped in the folded end. I like to think that I have that knack, that if there were an Olympic event for netting insects then I would be a strong contender. Some insects are flightier and faster than others, and each requires a different approach. Bees are actually amongst the easiest of insects to catch because they are focused on the job of visiting flowers and will let you approach very close. Also, once their head is inside a flower they cannot see anything for a moment or two. I usually choose this moment to move close and then wait for them to drop backwards off the flower, at which point they hover briefly while choosing which flower to visit next. A quick sideways swipe and a flick of the wrist and the bee should be safely in the bag. On this occasion I fluffed it. Maybe it was jet lag, but instead of calmly approaching, I leapt forwards to strike like a total novice, lost my footing on the shingle bank and fell head first into the bugloss bush on which she was feeding. Before I could extricate myself she was gone.

Mick had wandered a way off into the shrubs, and when I shouted he ambled over and unhelpfully observed that I was an idiot. We scoured the flower patches for the rest of the day but could find no more.

Rumour had it that the best place to find short-haired bumbles was Twizel (sadly pronounced Twy-zel, not Twi-zel, which is somehow more amusing). It was 20 miles or so beyond Lake Tekapo, so after a night in a pleasant hostelry we drove on. Twizel is a popular base for skiing and is said to be lively in winter but in summer it is a very quiet town, one of those miles-from-anywhere places where it is hard to work out what anyone does for a living. It didn't seem to have anything like as many flowers as Lake Tekapo but we did find some promising patches along the Twizel River bank and the shores of Lake Ruataniwha. Search as we might, however, we could find no short-haired bumblebees. I began to wonder if I hadn't imagined the one at Tekapo. Maybe they had also gone extinct in New Zealand, and our plans for a reintroduction would have to be abandoned.

On our second day in Twizel we were driving through the back lanes of the town looking for more flower patches when we spotted a splash of colour off to our left. On investigation this turned out to be the town rubbish tip, a sprawling area of building rubble with a mountainous pile of broken bottles. Clearly the tip wasn't heavily used – perhaps they had started taking rubbish elsewhere – for lots of weeds had sprung up, including thistles, lupins and yet more viper's bugloss. We thought it would be worth a look and so started working our way through the treacherous terrain, climbing over rusting girders and crumbling piles of concrete, nets in hand. And here it was, on perhaps the least scenic spot in all of New Zealand, that I caught my first short-haired bumblebee. She was a plump worker, the brownish stripes on her abdomen giving her a slightly grubby appearance in flight. If I'm honest, the short-haired bumblebee is not the most beautiful insect in the world. They have, as you might surmise, rather short hair. The females are mostly black with a number of yellowish-brown stripes, sometimes with a greenish tinge, and a scruffy white tail. But after flying 12,000 miles and searching for five days we were thrilled. In fact the Twizel tip turned out to be a short-haired bumblebee hotspot. We saw five more before the day was out. That night we celebrated with a couple of local beers and an especially large pie each.*

We spent the next few weeks searching central South Island

* Their delicious pies became something of a daily treat for us in New Zealand – I can particularly recommend the venison pies in Arrowtown if you happen to be passing, but you might want to skip the possum pies of Hokitika.

for short-haired bumblebees, travelling down to Queenstown in the south and as far as Palmerston on the east coast. They were not common anywhere, but we found quite a few. We mapped wherever we found them, and made notes as to which flowers they visited. To start with we used a GPS to map each bee's position – that was until I accidentally left the GPS on the roof of the car and drove off. We also watched each bee closely to see whether it was collecting pollen or nectar on each flower – nectar they drink with their long tongues, pollen they gather with their hairy legs and store in sticky balls in the pollen baskets on their hind legs. Short-haired bumblebees have long tongues, and so prefer deep flowers. They adore viper's bugloss for nectar, and also sometimes collect its purple pollen. Their favourite pollen source seems to be red clover, still often grown as a ley crop in New Zealand. We also saw them visiting bird's-foot trefoil, St John's wort and thistles, but they seemed to have a pretty narrow diet. Try as we might we could find no nests, although we were struck by the fact that almost every short-haired bumblebee we found was close to a lake, and the shores of all the lakes have stony banks. Twizel tip also has great piles of stones. Perhaps they like to nest under these stones where they would be warmed by the sun.

We took genetic samples to test if they were inbred – a technique which involves snipping a 'toe' off the bee (not really a toe, strictly speaking the final tarsal segment). As you might imagine, they really don't like this, but experiments on more common bumblebee species have shown that it neither shortens their lifespan nor reduces their ability to gather food for the nest, so although we felt rather mean we were able to console ourselves that we weren't doing any real harm.

By the time we returned home to the UK (with a bag full of little pots containing pickled bumblebee toes), we felt that we had learned quite a bit about the short-haired bumblebee. So far as we could tell, it does not need anything particularly unusual to survive – lots of red clover and viper's bugloss would go a long way and it shouldn't be too difficult to create big patches of these kinds of flowers. A bigger question was how to get the bees back. Going in the other direction, the hibernating queens had been dug from the ground in autumn. The short-haired bumblebee must have been much more common in Kent in the 1880s than it is now in New Zealand because I think you could probably dig all winter long without finding a single one. We could catch queens in December and January in New Zealand as they came out of hibernation, but if we brought them back to the UK at that time of year there would be no flowers and they would die. Nor would it be possible to put them back into hibernation as they would have just woken from an eight-month sleep – they would never survive another six months and then have the energy to build a nest. Clearly some thought needed to be applied if we were ever to get them back living in the wild in Britain . . .

CHAPTER TWO
The Bumblebee Year

A bee is never as busy as it seems; it's just that it can't buzz any slower.

Kin Hubbard (American humourist)

Until the mid-1800s, so little was known about bumblebees that even the basics of their life history had yet to be described. Charles Darwin was one of the first to study bumblebees, but he had diverse interests ranging from barnacles to worms to coral reefs, and bumblebees could necessarily occupy only a small proportion of his attention. The first person to devote himself wholeheartedly to the study of bumblebees was Frederick William Lambart Sladen, the eldest of twelve children born to Lieutenant Colonel Joseph Sladen, commanding officer of the School of Gunnery at Woolwich. Born in 1876, Frederick spent his childhood at the family home, Ripple Court near Dover in Kent. Victorian Britain had a strong tradition of studies of natural history, particularly among those of 'independent means', the lucky few who could afford to spend their days skipping through flowery meadows, pinning butterflies and pressing flowers (or at least that is how I imagine it). Schooled at home by private tutors, Sladen seems to have had plenty of time for raking through the fields surrounding his home looking for bumblebee nests, which he dug up and brought home for study. Astonishingly, by the age of sixteen he was the world expert, publishing a short book on bumblebee habits in 1892, and continuing to study them for the next twenty years, culminating in the publication of *The Humble-bee; its life history and how to domesticate it* in 1912. This was the first proper book devoted to bumblebees, and it is still an invaluable guide, containing many fascinating observations, anecdotes and descriptions of the experiments that he carried out in his garden. Species that are today rare or extinct in Britain, such as the short-haired bumblebee, were familiar to Sladen, and his descriptions of the nests of such species remain pretty much all that we know. No one since has come close to matching Sladen's knowledge of the nesting habits of bumblebees. Much of the information that follows in this and subsequent chapters was first discovered by Sladen.*

The bumblebee year really begins with the emergence of the first brave queens from hibernation in spring, often as early as February or March. These queens are the biggest of their kind, and since only the very fattest survive hibernation, the first to emerge are zeppelins of the insect world. By this time they are nearly starved, having been in hibernation since the previous summer, eight months or more ago. They drone slowly through the cold air, labouring to find scarce spring flowers. Pussy willows are one of the few plants in flower this early, and these small trees can attract hundreds of hungry queens. The flowers droop under the weight of the enormous bees. Only the female trees produce

* It is lucky for us that Sladen was so precocious, as he didn't live long. In 1912, the year he published *The Humble-bee*, Sladen was offered and accepted a job as an entomologist in Canada, where he worked mainly on honeybees rather than bumblebees. Sadly, after a hot day's work with his bees on Duck Island in Lake Ontario, he took a dip in the lake to cool off, suffered a heart attack, and died at the age of forty-five.

sugar-rich nectar, and it is this that the queens need first to boost their energy levels after their long sleep.* By contrast, the male trees produce bright yellow pollen-bearing catkins, and this is where the queens go next to stock up on the protein-rich pollen. Inside their abdomens, their ovaries are shrivelled and need protein to expand and develop their eggs. Sperm is also stored inside each, from a brief tryst the summer before with a now long-dead male.

Over the next few weeks the queens slowly fatten, and as their eggs begin to develop they start to search for nest sites. Many species like to nest underground, but as bumblebees are pretty poor diggers, they often explore existing holes. This nest-searching behaviour is very characteristic, particularly in the buff-tailed bumblebee, one of our commonest and largest species. These huge bees sway from side to side just above the ground, quartering across grassy terrain and along hedge banks in search of a mouse hole or other cavity. They are quite easy to follow and watch. Any dark patch of soil or indentation in the ground will draw their attention, and they will land to investigate. If there is a hole, they will crawl inside. Often they remain underground for many minutes, exploring subterranean chambers and tunnels dug by rodents or moles or rabbits. They may walk for many feet underground (some nests can be 10 feet or more along tunnels). Exactly what they look for we do not know, but presumably they are trying to assess whether the cavity has all the attributes required for a successful nest. Is it large enough? Is it deep enough to be safe

* Gender in plants is a much more complicated business than in most animals. A few plants such as pussy willows and red campions are either male or female. However, most plants are hermaphrodites, having both male and female reproductive organs in the same flower, which poses the danger that they might accidentally mate with themselves.

from badgers? Will it flood in a rainstorm? Is it dry and free from draughts? Is it already occupied? More often than not they eventually emerge and fly away to try elsewhere, but they must find somewhere or die in the search.

Although bumblebees do not gather nest material from far away like birds, they do need insulation in the form of feathers, hair, dried moss or grass, so this probably plays a large part in their choice of nesting site. Very often they seem to choose places where there is an old bird, rabbit, mouse or vole dwelling. They will enthusiastically rearrange the materials that they find in and around the nest into a nice cosy ball with a hollow centre. Some bumblebees, known as carder bees, will even vigorously comb it with the bristles on their legs. Their name is derived from this, carding being an old-fashioned term for combing wool before spinning. Bumblebees will readily use man-made materials. Loft insulation seems to be just the job, and some bumblebee species nest in roof spaces under the insulation (although modern felted roofs are much harder for them to penetrate). I have even heard of a bumblebee nest in a disused tumble dryer, with the bees using the accumulated fluff in the air filter to make themselves very snug.

Just like birds, different bumblebee species tend to nest in different places. White-tailed bumblebees seem to love settling under the wooden floors of garden sheds, for instance, whereas buff-tailed bumblebees often end up under patios, or will use airbricks to access the wall cavities of houses. Early bumblebees will sometimes nest in old birds' nests high in trees, and both the early and the red-tailed bumblebees will use tit boxes, but only if they contain an old bird's nest for them to use as insulation. Queens of the feisty Turkish bumblebee species, *Bombus niveatus,* can be so bold as to take over occupied redstart nests, driving the birds away even though the redstarts are perhaps fifty times heavier than them. The tree bumblebee, as its name suggests, always nests in holes in tree trunks, and again will often use tit boxes, whereas the carder bumblebees tend to nest just above the soil, in dense tussocks of grass or under piles of dry leaves under bramble thickets. Sadly, one of the few places where bumblebees generally won't settle is in the bumblebee nest boxes widely sold in garden centres.

Whatever site she chooses, within it the queen bee constructs a loose ball of insulating material with a central cavity and a hole through which she can squeeze. Within this cavity, perhaps the size of a tennis ball, she constructs a thimble-shaped cup from wax produced from special glands on the underside of her body, which is then scraped off and moulded into shape with her legs and mandibles. She fills this cup with honey (honey being concentrated nectar). She also collects pollen, which she forms into a small ball about the size of a pea, using a drop of sticky honey to hold it together. This she also coats in wax. By this time her ovaries should have fully developed, and she is ready to lay eggs, which she fertilises as they pass out from her body, using the sperm stored inside her. Next, she excavates a small hole in her pollen ball, pushes a batch of sixteen eggs into the dough-like material, and then seals them over with wax. The number sixteen is determined by the paired ovaries, each of which can produce eight eggs at a time. The eggs themselves are cream-coloured and shaped like slender sausages.

The mother incubates her laid eggs in much the same way as a bird. She shapes the pollen ball to create a shallow groove along the top into which her body fits snugly. She has thin fur on her belly so there is good contact between her and the brood (just like the brood patch of a bird). Now she begins to shiver, and warms the eggs, keeping them at about 30°C even when the air temperature outside the nest may dip well below freezing on an early spring night. Shivering uses lots of energy, which is why the queen has already placed her pot of honey within easy reach while she sits on the eggs; but this is not enough to keep her going for long. If she leaves her eggs for too long to collect more nectar they will get cold, but if she does not go to fetch food she will starve. A queen may use her own weight in sugar each day to incubate her brood, which may necessitate visiting up to 6,000 flowers. If these flowers are too few and far between she will be away from the nest for much of the day, her brood will cool and as a result develop too slowly, and she will wear herself out in her frantic search for food. Hence the proximity of lots of nectar-rich spring flowers is probably vital.

All being well, after about four days the eggs hatch into tiny comma-shaped white grubs. They have no legs, and are little more than eating machines, with a mouth at one end and an anus at the other. These unattractive creatures remain clustered together, a herd of maggots grazing on the pollen dough. As they use it up, the queen must gather more by visiting flowers, and add it to the brood clump.

To grow, the grubs must also shed their skins. This is a general feature of insect development – namely that their skin cannot grow with them as ours does, but is instead a more or less fixed size. So for the grub to get larger, the old skin must be sloughed off, revealing a new and larger skin underneath. Think of it as a paper bag – after each moult, this new skin will be half-full and wrinkled. As the grub feeds, this skin stretches out until the wrinkles are gone. At this point it cannot grow any larger, for if it eats more, its skin will split – which is exactly what happens. The grubs will shed their skins three times in about two weeks, by the end of which they will have increased enormously in size and will each be roughly the size of a broad bean (although they vary a lot). At this point the grubs each spin themselves a neat egg-shaped cocoon, using silk produced by glands in their mouths. Once inside their cocoon, they shed their skin again, but what emerges is very different: now the pupa has the rough shape of an adult bee, but retains the creamy colour of the grub. The legs, head, antennae, eyes and tongue are clearly visible, and neatly folded against the body. Inside the pupal skin an amazing transformation is taking place. The internal structures of the grub – the nerves, muscles, gut and so on – simply dissolve, and the body rebuilds itself in the entirely different form of an adult bee.

This reconstruction takes about two weeks, at which point the adult bee bursts out from the pupal skin and bites through the cocoon to escape. The first bees to emerge are all females, all the daughters of the queen; they are her workers. When they first emerge they are entirely white, resembling, with a little imagination, rather cuddly miniature polar bears. The wings are initially shrivelled, but in the first few minutes after the bees emerge, blood pumps into their veins to stretch the wings out. Once expanded, they quickly harden. The white appearance lasts for a couple of days, during which time the workers remain within the nest and the queen makes her last few foraging trips. As soon as her first batch of larvae has pupated, however, the queen will have laid a second batch of eggs and by now these should be well-developed grubs. The first generation of grown-up workers now takes over the care of these grubs, helping to keep them warm and feeding them. Once their adult colours have developed, some of these workers venture outside to gather food. At this point the queen stops the risky business of gathering her own food, and from here on she remains in the nest for the rest of her days, being fed by her worker children.

The workers are initially tentative when they leave the nest. Their first flights are short, as they familiarise themselves with the entrance to the nest and surrounding landmarks; if they become lost, then they are doomed. After a few exploratory trips they start to range further, and experiment with visiting flowers. They have innate preferences for blues and yellows, so will tend to visit flowers of these colours first. Finding nectar and pollen in flowers, and learning to gather it efficiently, is harder than it might at first appear, and it takes a few days for the worker bees to perfect their skills. Once they have, however, food begins to flow rapidly into the nest, and the queen starts to lay batches of eggs more regularly. If it is in a good position, with lots of flowers nearby, the nest grows rapidly. The workers also produce wax, like their mother, which they use to build more honey pots, and in some bumblebee species they also build special pollen-storage containers in the form of tall wax pots with a flared rim. Old pupal cells, once their occupants have departed, are remodelled with wax and used for honey storage. Often a wax dome is also built over the nest, which helps to insulate it and keep predators out. Compared to the military precision of a honeybee nest, which contains regular flat sheets of perfectly hexagonal cells, the bumblebee nest is a rather ramshackle, jerry-built affair, but it serves well enough.

By July, the nest may have grown to contain a sisterhood of several hundred worker bees, labouring together to help their mother. In physical size, the nest may now be as big as a half-deflated football, although this varies and some species, such as the early bumblebee, are content with a much smaller nest, more like a tennis ball. At this point in the summer, the queen changes strategy and ceases to produce daughter workers. Up until now she has been producing a chemical signal, a pheromone, which tells the developing grubs to become workers. Now, she ceases producing the pheromone and starts to lay both male and female eggs.* In the absence of the pheromone signal, these female grubs

* As will be explained later, a quirk of the genetics of bumblebees means that it is very easy for the queen to control the sex of her offspring by laying either unfertilised eggs, which become sons, or fertilised eggs, which become daughters. It is probably a good thing that humans are unable to do this.

develop as future queens, growing much larger than the grubs destined to become workers, and so take longer to reach full size. The males are smaller, similar in size to the workers. A big nest might produce a hundred or more future queens, and several hundred males, although most nests seem to produce either mainly new queens or mainly males.

Once they have emerged from their pupae, these new males and queens spend a few days resting, feeding on honey and pollen reserves. They will sometimes go out of the nest on brief exploratory missions, and before long they leave the safety of the nest for ever. The males are doomed to be short-lived for, with summer's end approaching, they have no way of surviving the winter. Their only role is to mate. In high summer, males can be very common. They sit around on flowers drinking nectar; they prefer flowers with big, sturdy heads such as thistles and knapweeds, and gangs of males can often be seen clustered together, reminiscent of a group of men propping up the bar in a pub.

Mating in bumblebees is a rather enigmatic activity that we will return to later. Even for some very common bumblebee species we are not sure how males and queens find one another as mating is so rarely seen. So far as we can tell, virgin queens are soon accosted by males when they leave the nest, and they tend to mate very quickly. As they enter hibernation soon afterwards, few queens are seen in high summer. In most species, the queens mate only once in their entire lives, and the sperm from this event is stored within them for use the next spring when each builds her own nest. The males will happily mate many times, but because there are usually many more males than queens, it is a very lucky male indeed who mates more than once in his life.

Once mated, the queens must find somewhere to hibernate. This clearly doesn't take long as, by contrast to their nest-searching behaviour in the spring, they are rarely seen looking for hibernation sites. They hibernate a few centimetres under the ground, but as bumblebees are pretty inept at digging, they tend to choose very loose soil that is easy to burrow into. They will also sometimes go into newly dug soil in gardens, and I have a suspicion that in the countryside they often make use of molehills. Hibernating queens are also frequently found in old compost heaps and in flowerpots full of loose compost.

Back in the old nest, the founding queen is now a little over a year old. She is frail and balding. Her workforce is no longer being replaced, for all the colony's reserves have gone into producing new queens and males. As the ageing workers die off, the food supply coming in dries up, and before long all of the bees are dead. By September, all that remains is a pungent and loose pile of empty pupal cells, honeypots and bee corpses, which are slowly picked through and eaten by a range of scavengers such as wood-lice, maggots and beetles.

The new queens of the aptly named early bumblebee enter hibernation as early as June, while most species do so in July or August. We don't know why they hibernate so early, but perhaps hibernating is safer then being out and about, where the queen might pick up parasites or be eaten by a predator. They have a long wait until the following spring. The fat reserves in the queen's body must keep her going; any queens that are smaller than average or have slightly smaller stores of fat tend to die during the long hibernation. They are also prone to becoming mouldy in damp weather, or drowning in heavy winter rains, and it is likely that the vast majority of hibernating queens do not make it through to spring.

Fortunately some do, and as the first rays of spring sunshine warm the soil, they burrow upwards to the light, beginning the bumblebee year once again.

CHAPTER THREE
The Hot-blooded Bumblebee

Aerodynamically, the bumble bee shouldn't be able to fly, but the bumble bee doesn't know it so it goes on flying anyway.

Mary Kay Ash
(American businesswoman and writer)

By 1974 I was nine years old. I became interested in how animals worked, and decided to investigate what they looked like on the inside. Thinking back, I wonder at my parents' indulgence, for I took to collecting roadkill and dissecting the remains of a variety of mangled beasts in our back garden. Rabbits were the most common, but I also found grey squirrels, a fox, a lovely brown hare, pigeons, frogs and who knows what else. I persuaded my grandparents on my mother's side to buy me a dissecting kit for my birthday. I've no idea now what I said I wanted it for, but I must have made up some sort of plausible excuse. I very much doubt that I told them I wanted to chop up dead and squashed animals. They were very strait-laced, devout Methodists from a small village in deepest, darkest Norfolk, and not exactly humorous. I can't imagine that they would have thought this a healthy hobby for a nine-year-old. Nonetheless, on my birthday the dissecting kit arrived (ordered from Watkins & Doncaster, of course), and it was marvellous, containing lethally sharp scalpels of various shapes and sizes, a selection of pointed probes, delicate scissors and a lovely little chrome hand lens, all rolled up in a soft cloth pouch. I boosted its contents by persuading my uncle Ed, a chiropodist from Norwich, to give me some extra instruments including some wickedly curved scalpel blades used for carving ingrowing toenails or bunions or some such.

Fully equipped, I would slice the bellies of the unfortunate animals open and carefully remove their innards, piece by piece. My mother had a copy of *Gray's Anatomy* from her college days, and although it describes human anatomy, I used it to identify the various organs, many of which look surprisingly similar in small mammals and even in birds. I remember being struck by the fact that the kidneys of rats were exactly the size and colour of kidney beans. I would lay the organs out on the slabs of the garden path, and then puzzle over the pieces that I could not identify. I was never very good with spleens. I managed to get hold of some formaldehyde, and used it to pickle some of the more interesting body parts in jam jars, which I arranged on the shelves in my bedroom.

From dissection, it was a short but inevitable move to taxidermy. I have already mentioned that the Watkins & Doncaster catalogue contained a selection of tools and chemicals for taxidermy, including glass eyeballs of every possible size and colour. I now managed to get hold of a book which described how to do it – the wonderful *Home Book of Taxidermy and Tanning* by Gerald J. Grantz, which I still have to this day, the pages stained from various bits of animal innards. I saved up and bought the necessary chemicals and some taxidermy needles; curved along their length and of triangular cross section, they slide through the hide of an animal much more easily than a normal sewing needle.

Of course roadkills are rarely of much use for taxidermy: if the body is squashed, split and covered in tyre marks it is hard to reconstruct to a lifelike form. Getting hold of undamaged but dead creatures was something of an obstacle. My first specimen was a black-headed gull which I found on a day out to Ellesmere, a small and pretty town with a large lake and many waterfowl. Feeding bread to the ducks, I noticed one particular gull which was unable to fly and seemed quite sick, so I caught it, took it home and attempted to nurse it back to health. By dawn it was dead. Since it was unmarked, it seemed a perfect candidate for my first taxidermy attempt.

The principles of taxidermy are very simple. The body is sliced open in a neat line running from the bottom of the tail to the base of the neck. The skin is then peeled back and the carcass pulled out, turning the skin inside out. In a bird the wing bones and leg bones are chopped through, leaving the wings and feet attached to the skin. The neck has to be severed, leaving the bulk of the carcass free from the skin. The skin is then peeled forwards over the skull to the beak, carefully cutting around the eyes. The tissue, eyeballs, brains and other soft bits are then removed from the skull as best one can. In mammals, because there's no beak attached, the skull can be completely removed and boiled in a pot – it is then much easier to get the brains out. The skin is treated with borax to preserve it, and left for a day or two. In the meantime, one has to construct a body out of wood wool (fine wood shavings) bound with lots of cotton thread. The animal's carcass is used as a model, and as you might imagine it is important to get the shape as close to the original as possible. Three stiff wires (Mother's cannibalised coat hangers) are then pushed through the body. One protrudes on either side, for attachment to the wing bones or forelegs. The second is sharply curved and pushed into the underside so that both ends point downwards, for attachment to the back legs. The final one is shorter, and is pushed into the top of the body where the neck would naturally attach; only one end is left protruding, on to which the skull is impaled. The skin is then rolled back over the skull, stretched over the new body, and the legs and wing bones attached to the appropriate wire with more cotton. The skin is sewn back up and *voilà*! One has a perfect replica of the original animal. Or at least that is the theory.

In practice, there were many problems. Professional taxidermists keep a range of eyeballs ready for use. I could not afford to do this, so I had to buy them after a suitable corpse had presented itself, and they often did not arrive until a few weeks later.* Corpses don't keep well (my mother wouldn't let me put them in the freezer), so I had to stuff them without the eyes, then add the eyes some time afterwards. This is not ideal. Normally, the glass eyes should be placed in the skull before the skin is pulled back over it, but this was not possible, so I had to try push the eyes in through the eyelids from the outside once they arrived. By this time the skin had often hardened and shrunk a little, so the new eyes rarely sat comfortably; usually they were left protruding somewhat, giving my stuffed animals a startled expression, rather as if something had just been shoved up their bottom (which of course wasn't far from the truth). Moreover, the rapidly shrinking skin would often not fit back over the new body, leaving a bit of a gap at the front from which wood shavings protruded. I eventually learned to compensate for this by making the wood-wool bodies smaller than the originals, but this gave my animals an emaciated appearance. I also found that it was surprisingly difficult to arrange the legs and wings in a pose that looked remotely lifelike. In my gull, the wings slanted upwards at an awkward angle, one leg was stubbornly turned sideways, and various feathers stuck up at unexpected angles. The overall effect was of a bird that had just received a large electric shock. Nonetheless I proudly took it in to my primary school class art exhibition. Dear Miss Scott remembers the gull to this day, although she must now be well into her eighties.

My efforts to stuff other animals were no more successful. I found a ferret dead in a snare, and deployed my limited skills once

* Younger readers might struggle to imagine a world where orders were sent by post rather than placed online, and 28 days was considered a reasonable turnaround time for delivery.

more. This time, the skin shrank so badly that I had to make the body incredibly narrow to stretch the skin around it. The poor creature ended up so thin it looked as if it was constructed from pipe cleaners. It also smelled spectacularly awful; live ferrets smell bad enough, but a dead one defies description. It was soon consigned to the garage, by order of my mum.

I came across a sickly wood pigeon which, miraculously, I managed to nurse back to some semblance of health by giving it dog worming tablets. After a couple of days of rest and recuperation in a cardboard box with plentiful hamster food to eat, it seemed quite chirpy, so I deemed it ready to return to the wild. With a bit of encouragement it took off from my bedroom window, but being somewhat weakened by its recent illness it crash-landed 100 yards away in the field opposite. I scampered outside and across the road but as I was climbing over the wall I saw that I was too late. The bad-tempered horse had trotted over and stamped on the pigeon with a front hoof, undoing my good works and leaving me with a slightly squashed corpse. Whilst the back end was badly damaged, the front, with its beautiful iridescent neck feathers, was fine. My solution was to mount the head and neck on a small shield-shaped piece of plywood, and hang it on the wall, rather like the heads of deer which are traditionally displayed as hunting trophies (but somewhat smaller and less impressive). The result of this new obsession was that my bedroom became more cluttered than ever, adorned now with a myriad of nightmarishly deformed creatures.

In the meantime, my fascination with insects had not abated. I became much better at rescuing forlorn bumblebees. I had noticed that it is not uncommon to see bumblebees, particularly queen bees in spring, walking slowly along the ground – particularly obvious when they are walking along pavement. I found that if I put my hand near them, they would feebly raise a middle leg, the defence posture of the tired bee; but that it was then possible gently to coax them up on to my hand and that they never stung. Having learned my lesson about warming them up using the cooker, I now tried a different approach and found that they would readily sip a mixture of honey and water in a teaspoon placed in front of them, and that after half an hour or so they would often revive and fly away (although sometimes they would climb into the teaspoon and become hopelessly wet and sticky). It was a few years before I came to understand exactly what was going on here, and it is quite revealing about the biology of bumblebees.

When I later went to secondary school, I loved biology lessons; it was, you will not be surprised to hear, my favourite subject. My biology teacher there was a short, rotund and enormously hairy man called Mr Newton (predictably known to us as Isaac) with a huge Sherlock Holmes-style pipe permanently clamped between his teeth. Reeking of Clan, his favourite tobacco brand, he taught us that insects, fish, amphibians and reptiles are cold-blooded, while mammals and birds are warm-blooded. Now Isaac was an excellent if rather grumpy teacher, but he had got this wrong, although he had no way of knowing it. In fact, at almost exactly the time I was being taught this in 1976, an American scientist called Bernd Heinrich* was making a name

* In addition to being a ground-breaking and prodigiously productive scientist, Bernd Heinrich was a phenomenal marathon runner, with a best time of 2 hours, 22 minutes and 34 seconds, only narrowly missing the US Olympic team at the age of forty. He went on to set the American record for 100 miles, for 100 kilometres, and for the furthest distance run on a track in 24 hours (156 miles, 1,388 yards!). Quite how anyone could face running around a track this many times is something of a mystery to me. However, I have recently taken up marathon running myself and have found that I am quite good at it (best time about 3 hours so far), so perhaps there is something about studying bumblebees which makes one good at long-distance running.

for himself by stabbing bees and hawkmoths. He was not stabbing them for fun, but rather to take their temperature. Measuring the temperature of a bee is quite tricky, as neither their mouth nor their bottom can accommodate a conventional thermometer. Instead, Heinrich used a thermocouple, a needle made from two thin wires of different metals welded together, attached to an electrical meter. As the temperature of the thermocouple varies, so the electrical conductivity of the junction between the metals changes. By stabbing the needle into a live bee, its temperature can be measured from the conductivity reading. Of course this isn't much fun for the bee.

Heinrich developed a manic curiosity for spearing bumblebees and other large insects such as hawkmoths and dragonflies, and he found that these big, fast-flying insects are far from cold-blooded (his discovery coming at substantial cost to the local insect population). In fact, flying bumblebees have a body temperature that is generally well above that of the air around them, and tends to be constant at about 35°C, close to the usual temperature of a human body. An even rudimentary grasp of physics suggests that this is quite extraordinary. Keeping warm is harder the smaller you are. Big animals, such as blue whales, have a small surface relative to their volume, and so they cool very slowly and can keep warm even in very cold conditions (such as in the Antarctic Ocean). In contrast, small creatures, such as flies, have a vast surface area relative to their volume, and lose heat incredibly quickly. Yet bumblebees, which in the grand scheme of things are rather closer to flies than to blue whales, can keep themselves warm even when the surrounding air is 30°C cooler than their body temperature; a phenomenal feat. How do they do this?

Heinrich found that the answer is in two parts: keeping heat in, and generating it in the first place. Keeping warm is helped if you have insulation, and of course bumblebees have furry coats. Some bumblebees which live in the Arctic have particularly long fur, and they also tend to be bigger than more southerly bumblebees, which helps. The vital thing for a bumblebee is to keep its thorax (the middle section) warm, because this is where the flight muscles are; unless the thorax is warm enough, the muscles cannot contract sufficiently fast, and the bee cannot take off. The temperature of the abdomen (the hindmost section) doesn't matter much in flight. The abdomen and thorax in a bumblebee are connected by a very narrow waist, and the front part of the abdomen contains a sac of air, so heat loss from the thorax to the abdomen is minimal (air is a poor conductor). Heinrich found that the abdomen in a flying bee could often be 15°C cooler than the thorax.

A furry coat and insulating air bags are a good start, but the heat has to come from somewhere, and it is generated by the contractions of the flight muscles. In flight a bumblebee flaps its wings 200 times per second (which equals 12,000 rpm), roughly equivalent to the speed of a high-revving motorbike engine. This generates a lot of heat, but of course comes at a cost: bumblebee flight is enormously expensive in terms of the energy that it uses. Many of the details have recently been worked out by Charles Ellington's team at Cambridge University. They persuaded bumblebees to fly in a sealed wind tunnel, within which they were able to measure the bee's oxygen usage and hence calculate its metabolic rate. Persuading a bee to fly in a sealed chamber is pretty tricky. Place her in a jar and she will take off and buzz up and down the side for a while, but it is not very much like natural flight. Creating wind using a fan in the chamber achieves very little; the bee either sits tight on the floor while the wind whistles past or she takes off but immediately crashes into the side of the chamber and falls to the bottom. Neither is very helpful. The secret to persuading her to fly for prolonged periods is to create a moving landscape that rushes past her as if she were flying along. This can be made by painting a pattern on to a loop of material stretched over a pair of motorised rollers. This contraption is then placed beneath the glass chamber. As the rollers turn, the landscape appears to move, which in combination with the wind convinces the bee that she is making good progress, even though she is actually going nowhere.

Using this set up, Ellington's team could measure the amount of oxygen used up by each flying bee. This in turn enabled them to calculate how much energy bees burn in flight: an estimate of about 1.2 kJh^{-1}. That figure may not mean a lot, so let me contextualise by saying that a running man uses up the calories in a Mars bar in about one hour. A man-sized bumblebee (which would, I admit, be pretty terrifying) would exhaust the same calories in less than thirty seconds. Hummingbirds are often thought of as having exceptionally high metabolic rates, but a bumblebee's is roughly 75 per cent higher.

This simple fact explains an awful lot about the biology and conservation of bumblebees. They have to eat almost continually to keep warm; a bumblebee with a full stomach is only ever about forty minutes from starvation. If a bumblebee runs out of energy, she cannot fly, and if she cannot fly, she cannot get to flowers to get more food, so she is doomed – unless a small boy comes along and gives her a teaspoon of honey. With a stomach full of sugar she can start to fire up her flight muscles, shivering them to produce heat, and once she gets up to about 30°C, off she goes . . .

A bumblebee's dense furry coat has obvious advantages, but it can create problems when the weather is warm. Bumblebees cannot help but produce lots of heat when they fly, which can be difficult to get rid of if the air temperature is also high. This is probably why bumblebees are not common in Mediterranean countries, and why there are almost none in the tropics. If their body temperature exceeds 44°C they will die; as they approach this lethal limit their metabolism collapses and they become unable to fly. It is noticeable that those species of bumblebee that occur in warmer climates tend to have shorter fur, while those from high latitudes and altitudes tend to be very large and very furry. Hence the huge size and long shaggy coat of the world's largest bumblebee, *Bombus dahlbomii*, which inhabits the high Andes of South America.

Bumblebees do have a trick to help them lose heat in hot weather. As already mentioned, they normally keep their abdomen at a much lower temperature than their thorax, with the narrow waist connecting the two acting as a barrier. If the thorax starts to get too hot, the abdomen starts rhythmically to contract, sending surges of cool blood into the thorax and sucking back waves of hot blood. This heats the abdomen and so raises the surface area from which heat can be lost. Nonetheless, on very hot days in summer bumblebees will tend to stop foraging as noon approaches and recommence in the cool of evening.

Pumping heat from the thorax to the abdomen can also serve a very different purpose. The northernmost social insect in the world is a bumblebee known as *Bombus polaris*, which lives well within the Arctic Circle: large and unusually hairy, it can exist in regions where, even in the height of summer, air temperatures rarely exceed 5°C. Unlike other bumblebees, the *Bombus polaris* queens maintain a stable, high abdominal temperature (greater than 30°C) by pumping hot blood from their thorax to the abdomen. This enables them to develop eggs within their ovaries quickly, which is important in the very short Arctic summer. As the workers and males have no eggs to develop, their abdomens are substantially cooler.

The larger the bee, the easier it is for it to keep warm but the more prone it is to overheating in hot weather. This may explain why queen bees are so much larger than workers, for they are on the wing earlier, in the spring when the weather is cold. It may also explain why the worker bees that leave the nest to gather food tend to be larger, on average, than those that stay in the nest to look after the brood.

Moreover, bumblebees manage not only their own body temperatures, but also those of their brood and the nest. Dependent upon species, bumblebees have anywhere between two and seven months to complete their annual cycle. This would not be possible unless they speeded up the development of their offspring by keeping them warm, and as the grubs are flabby near-immobile creatures with very small muscles, they cannot warm themselves. Instead they are incubated by the queen (if they are the first batch of offspring), or by the workers. Once there are enough of them, the combined heat emanating from the workforce keeps the whole nest at a cosy 30°C or so without much difficulty. As with individual bees, overheating of the nest can be a problem in warm weather. If the nest becomes too warm, or if carbon dioxide levels climb too high, some workers will station themselves in the entrance and fan hot air out of the nest, acting like miniature air-conditioning units. Different bees have different temperature thresholds for beginning fanning behaviour; if the nest is slightly too hot, only one or two bees fan. If the temperature continues to rise, more and more join in. This very simple mechanism enables large nests to regulate their temperature very precisely, keeping it to within 1°C of 30°C day and night.

The ability of established bumblebee colonies to keep warm is most impressive. I once was looking for the kindest way to dispose of a colony of Turkish buff-tailed bumblebees – factory-reared bees which could not be released into the wild in the UK as they are not native here – and decided that freezing them was probably the best option. I placed the nest in its entirety in a domestic freezer at -30°C. The next day I came back to find the colony very much alive and buzzing loudly; the workers had gathered into a tight clump over the brood and were presumably shivering at maximum capacity. The queen was hidden in their centre, and seemed quite unperturbed.

CHAPTER FOUR

A Brief History of Bees

Let us travel back in time 135 million years. The vast super-continent of Gondwana was beginning to break up, with South America drifting off to the west of Africa, and Australia moving majestically off to the east. Antarctica decided to head south, dooming all but the most adaptable of its inhabitants to an eventual icy grave. The South Atlantic and Indian Oceans were slowly forming.

At this ancient time, an era known to geologists as the Cretaceous, the continents were clothed in green forests of tree ferns, cycads, huge horsetails, and conifers such as pines and cedars. This was the height of the reign of the dinosaurs, although not the species that are so well known to schoolchildren the world over: amongst the trees, herds of vast herbivores such as Iguanodon grazed, standing on their hind legs to reach higher foliage; heavily armoured, tank-like species such as Gastonia bulldozed through the undergrowth; and packs of ferocious meat-eaters such as Utahraptor hunted their prey. The air swarmed with primitive insects including oversized dragonflies and early butterflies, and this was also the heyday of the pterosaurs, the largest animals ever to fly above the earth, with wingspans up to 12 metres. Much smaller dinosaurs had also taken to the air; feathers, probably first evolved to help these little creatures keep warm, became elongated on their forelegs to allow gliding and, eventually, active flight. These were the first birds. Our own ancestors at this time were rather unimpressively small, rat-like creatures skulking in the undergrowth, nervously coming out at night to nibble on insects, seeds and fallen fruit. If we could travel to this ancient land, we might be too concerned with the dangers posed by the larger wildlife to notice that there were no flowers; no orchids, buttercups or daisies, no cherry blossoms, no foxgloves in the wooded glades. And no matter how hard we listened, we would not hear the distinctive drone of bees. But all that was about to change.

Sex has always been difficult for plants, because they cannot move. If one cannot move, then finding a suitable partner and exchanging sex cells with them poses something of an obstacle. The plant equivalent of sperm is pollen, and the challenge facing a plant is how to get its pollen to the female reproductive parts of another plant; not easy if one is rooted to the ground. The early solution, and one still used by some plants to this day, is to use the wind. One hundred and thirty-five million years ago almost all plants scattered their pollen on the wind and hoped against hope that a tiny proportion of it would, by chance, land on a female flower. This is, as you might imagine, a very inefficient and wasteful system, with perhaps 99.99 per cent of the pollen going to waste – falling on the ground or blowing out to sea. As a result they had to produce an awful lot.

Nature abhors waste, and it was only a matter of time before the blind stumbling of evolution arrived at a better solution in the form of insects. Pollen is very nutritious. Some winged insects now began to feed upon it and before long some became specialists in eating pollen. Flying from plant to plant in search of their food, these insects accidentally carried pollen grains upon their bodies, trapped amongst hairs or in the joints between their segments. When the occasional pollen grain fell off the insect on to the female parts of a flower, that flower was pollinated, and so insects became the first pollinators, sex facilitators for plants. A mutualistic relationship had begun which was to change the appearance of the earth. Although much of the pollen was consumed by the insects, this was still a vast improvement for the plants compared to scattering their pollen to the wind.

To start with, insects had to seek out the unimpressive brown or green flowers amongst the surrounding foliage. It was now to the advantage of plants to advertise the location of their flowers, so that they could be more quickly found and to attract insects away from their competitors. So began the longest marketing campaign in history, with the early water lilies and magnolias the first plants to evolve petals, conspicuously white against the forests of green. The first pollinators may have been beetles, which many water lilies still rely on to this day. With this new reliable means of pollination, insect-pollinated plants became enormously successful and diversified. Different plants now began vying with one another for insect attention, evolving bright colours, patterns and elaborate shapes, and the land became clothed in flowers. In this battle to attract pollinators, some flowers evolved an additional weapon – they began producing sugar-rich nectar as an extra reward. As these plants proliferated, so the opportunities for insects to specialise grew, and butterflies and some flies evolved long, tubular mouthparts with which to suck up nectar. The most specialised and successful group to emerge were the bees, the masters of gathering nectar and pollen to this day.

All bees feed more or less exclusively on nectar and pollen throughout their lives. While many other insects such as butterflies and hoverflies feed on flowers as adults, very few do so as young too. Flowers are sparsely distributed in the environment, and immature insects cannot fly from one to another as only adult insects have wings. The innovation unique to bees is that the adult females gather the food for their offspring, so that their larvae do not need to move at all. The larval stage is maggot-like, legless and generally rather feeble, being defenceless and capable of only very limited movement. They are entirely dependent on the food provided by the adult bees.

The first bees evolved from wasps, which were and remain predators today. The word 'wasp' conjures up an image of the yellow-and-black insects that often build large nests in lofts and garden sheds and which can be exceedingly annoying in late summer when their booming populations and declining food supplies force them into houses and on to our picnic tables. Actually, there are enormous numbers of wasp species, most of whom are nothing like this. A great many are parasitoids, with a gruesome lifestyle from which the sci-fi film *Alien* surely took its inspiration. The female of these wasps lays her eggs inside other insects, injecting them through a sharply pointed egg-laying tube. Once hatched, the grubs consume their hosts from the inside out, eventually bursting out of the dying bodies to form their pupae. Other wasp species catch prey and feed them to their grubs in small nests, and it is from one such wasp family, the Sphecidae, that bees evolved. In the Sphecidae the female wasps stock a nest, usually an underground burrow, with the corpses, or the paralysed but still living bodies, of their preferred prey. They attack a broad range of insects and spiders, with different wasp species preferring aphids, grasshoppers or beetles. At some point a species of sphecid wasp experimented with stocking its nest with pollen instead of dead insects. This could have been a gradual process, with the wasp initially adding just a little pollen to the nest provisions. As pollen is rich in protein, it would have provided a good nutritional supplement, particularly at times when prey was scarce. When the wasp eventually evolved to feed its offspring purely on pollen, it had become the first bee.

Exactly how long ago this happened we do not know for insects rarely form fossils, and so we have to piece together their history from sparse information. Occasionally, insects become trapped in tree resin which fossilises to amber, beautifully preserving them for eternity. Crawling insects such as ants seem to have become trapped most often, but it seems that bees were rarely so foolish and examples of bee fossils are particularly few. The oldest known bee in amber is about 80 million years old, and is of a type known as a stingless bee, similar to species that live today in South America. These are advanced social bees that live in vast colonies, so it is a pretty good guess that the earliest bees were on the wing long before this.

A rather different source of information on the evolution of insects is provided by analysis of DNA sequences, which allow us to make educated guesses as to how long ago different evolutionary lineages diverged. Studies of the similarity of the DNA in wasps and bees suggest that the first bees appeared about 130 million years ago, 50 million years before the first known fossil bee, and probably very shortly after the first flowers evolved in the Cretaceous.

Over the millennia, bees have adapted to feeding on flowers in various ways. Many species have become hairy, which helps them to brush pollen from flowers, and also to hold it in flight. In the leafcutter bees, for instance, the pollen is stored among dense hairs on the underside of the abdomen, so that the bees often appear to have bright yellow bellies. In bumblebees and honeybees, stiff bristles on the hind legs form a basket into which pollen is placed. If one is going to visit flowers for their pollen it makes sense to also collect their nectar, for this is a great source of sugar to sustain flight. Nectar is expensive for plants to produce, and therefore many flowers evolved over time to hide their nectar, ensuring that only the insects most likely to provide them with a reliable pollen delivery service can reach it. Many bees evolved longer and longer tongues to make it easier for them to reach nectar hidden within flowers; some now have tongues longer than their bodies.*

The earliest bees, 130 million years ago, were almost certainly solitary species, and the majority of present-day bee species remain so. Each female builds her own nest, usually in a small hole in the ground, or in a tree or wall. In the leafcutter bees, the nest is lined with neatly snipped semicircles of leaves, glued together with silk. Once the nest is complete, the female bee fills it with pollen mixed with nectar and lays one or more eggs. The life cycles are very variable, but usually the female does not care further for her offspring, simply sealing up the nest entrance and leaving them to eat their pollen and develop on their own. Most solitary bees in temperate climates have just one generation a year, so the offspring will sometimes spend eleven months developing in the nest before emerging as adults.

Solitary bee species tend to be small, dark or drably coloured, which is why people seldom notice them. Nonetheless many are quite common and often live in gardens, some even nesting in the old mortar between the bricks of our houses. Only rarely do the lives of these inconspicuous creatures impinge noticeably on our

* The record holder is not a bee, but a hawkmoth, *Xanthopan morganii,* which has a tongue of about 30 centimetres long (the moth itself being 6 centimetres long). This moth feeds upon the Madagascar star orchid *Angraecum sesquipedale*, in which nectar is hidden at the base of spurs 30 centimetres deep, in a beautiful example of co-evolution. Upon being sent examples of the orchid in 1862, Charles Darwin predicted that there must exist a moth with a tongue long enough to feed upon it, but it was not until 1903 that the moth was finally discovered.

own, although they probably contribute substantially to pollination of many crops without us being aware of it (honeybees often get all the credit).

I was once involved in a rather strange and less welcome instance of a solitary bee impacting on humans. I received a call from aeronautical engineers who were investigating the cause of an instrument failure which had forced a military helicopter belonging to a certain well-known superpower – confidentiality agreements prevent me from revealing which one – to perform an emergency landing. A small but vital instrument which measures airspeed and controls the speed of rotation of the rear rotor had failed, and the British manufacturers of the instrument found themselves under suspicion of supplying dangerously defective components. Upon close examination, it transpired that the cause of the fault was a plug of a sticky yellow substance blocking a tiny but necessary hole in the instrument casing. Their investigations suggested that the substance might be pollen, which was when I was brought in. It was indeed pollen, identifiable as belonging to some species of legume, no doubt placed there by a small solitary bee which had adopted the hole as its nest while the aircraft was parked. When it returned from a foraging trip, the bee was presumably rather disappointed to find that its nest had vanished.

Let us return to our journey through time. To recap, bees first appeared perhaps 130 million years ago, and by 80 million years ago some had evolved a social lifestyle, for the earliest fossil is of a social stingless bee. Some 65 million years after the first bees appeared (and, coincidentally, 65 million years before the present), the earth went through a catastrophic change. Most scientists these days agree that a meteor struck the earth roughly where the Yucatan Peninsula now lies, causing tidal waves and massive volcanic eruptions which filled the air with so much dust that it blocked out the sunlight, in turn causing temperatures to fall below freezing for months or years on end. Almost all large forms of life on earth then died out very swiftly, the dinosaurs among them. Amazingly, representatives of many of the smaller groups of organisms survived somehow. So far as the sparse fossil record reveals, the main insect groups – bees, ants, grasshoppers, beetles and so on – seem to have recovered swiftly, although it is likely that countless individual insect species became extinct. The flowering plants also survived, presumably as dormant seeds. Our own ancestors – small, furry and warm-blooded – may have kept themselves alive by feeding on the corpses of larger animals or on stores of seeds and nuts, and perhaps by keeping warm in the vast drifts of rotting vegetation that resulted from the forests' death. Before long the earth was once again teeming with life, albeit with rather smaller forms.

Our mammalian ancestors took advantage of the many unoccupied niches and diversified. Were it not for the meteor, it is doubtful if most of the larger mammals – including ourselves – would ever have appeared. Some species grew much larger, filling the roles once occupied by dinosaurs; these included ground sloths that stood 6 metres tall and weighed 3 tonnes, and the vast rhinoceros-like *Uintatherium*. It was into this world of giants that the first bumblebees appeared, about 30 to 40 million years ago. This corresponded with a period of cooler temperatures, which may have encouraged bees to become larger and furrier. Our best guess is that the first bumblebee lived somewhere in the mountains of central Asia, since this is still the area of greatest bumblebee diversity. From here they spread west, east and north from the Himalayas to occupy Europe, China and Siberia, and even up into the Arctic Circle. As bumblebees overheat in warm climates, they did not spread far southwards towards the equator, which is why until some recent deliberate introductions there were no bumblebees in Australia, New Zealand or Africa south of the Sahara. About 20 million years ago bumblebees crossed from Siberia to North America, where they thrived and spread southwards. Eventually about 4 million years ago a handful of species moved down through the mountain chains of Central America to occupy South America, becoming the only naturally occurring bumblebees in the southern hemisphere.

So now we arrive at the present day. The world is blessed with an extraordinary diversity of species of organism. About 1.4 million have been named so far, but estimates as to the true total vary hugely from 2 million to 100 million. Two hundred and fifty of the known species are bumblebees (members of the genus *Bombus*, of which twenty-seven occur naturally in the UK). There may be a few more yet to be found in remote regions, but probably not many. There are about 25,000 known species of bee (superfamily *Apoidea*, with 253 known from the UK), but many more undoubtedly remain to be discovered, particularly in the tropical regions. Bees in turn belong to the immensely successful insect order the Hymenoptera, which also includes ants and the wasps from which bees evolved, of which there are 115,000 known species. The Hymenoptera in turn are just one of many types of insect, collectively the most successful group of organisms on earth, with about 1 million named species, or about 70 per cent of all known species on earth.

Until recently, this number of species was the highest it had ever been since life began. However, in the last few thousand years it has started to drop rapidly as man has remoulded the surface of the planet. As our ancestors spread out from Africa, many of the large mammals such as mammoths, giant sloths and sabre-toothed tigers swiftly disappeared, either hunted to extinction by man or driven to extinction because their prey disappeared. Most would have had no defence against groups of men hunting with spears and bows and arrows. At present, species are going extinct at somewhere between 100 and 1,000 times the natural rate, largely driven by habitat destruction and the ravages wrought by invasive species. It is estimated that one species goes extinct every twenty minutes.

So far, only three bumblebees are thought to have gone extinct globally: *Bombus rubriventris*, *Bombus melanopoda* and *Bombus franklini*, but surely more will follow. It is the threat of extinction of large mammals such as tigers or rhinoceros that tends to capture the public's attention, but arguably it is the loss of the smaller creatures that should give us most concern. Insects are responsible for delivering numerous 'ecosystem services' such as pollination and decomposition, and there is no doubt that little life on earth (including ourselves) could survive without them. As the famous biologist E. O. Wilson said, 'If all mankind were to disappear, the world would regenerate back to the rich state of equilibrium that existed ten thousand years ago. If insects were to vanish, the environment would collapse into chaos.'

CHAPTER FIVE

Finding the Way Home

Pigeons are not everyone's favourite creature. I must admit that I'm not enormously fond of the pestilential flocks of feral pigeons that infest many city centres, or even of the plump glossy wood pigeons that decimate my vegetable seedlings. Pigeons don't look particularly bright – in fact I've always felt that they have a rather vacant expression, and they do an awful lot of mindless cooing – but nonetheless they are capable of truly amazing feats of navigation.

Imagine this for a moment: you are locked in a dark box, transported for hours over 200 miles from home in a random, unknown direction, and then asked to find your own way back. You'd be rather annoyed and have no idea which way to go. You would of course ring the police, or ask someone for directions, but suppose you were unable to do either? Would you ever get home? Contrast this with the pigeon's response. With barely a moment's hesitation, the pigeon sets off flying at a brisk pace in precisely the correct direction. A few hours later it is happily perched in its loft tucking into a tasty bowl of grain. How on earth does it do this? Despite our vastly superior intellect and many years of scientific research, we have still not fully understood how pigeons navigate home so expertly from places that they have never previously visited. They can certainly use the sun or the stars as a compass, and can even tell where the sun is on heavily overcast days by their ability to detect the plane of polarised light penetrating the clouds. There is also some evidence that they have miniature magnets in their brains that enable them to detect the earth's magnetic field, so they have at least three inbuilt compasses. Impressive though all this is, you need more than a compass or three to find your way home if you have no idea which direction home is. And therein lies the mystery. It is almost as if they have a seventh sense (the sixth being the ability to detect magnetic fields) that we have yet to discover – a miniature GPS system perhaps, which tells them exactly where they are in relation to home.

At university I once had to write an essay on homing in pigeons, and became intrigued. In my later life as a university lecturer in the late 1990s, I couldn't resist investigating the navigation abilities of bumblebees. My experiments were very simple, and followed the model used by pigeon fanciers the world over. I set up five buff-tailed bumblebee nests in boxes in my garden. At that stage I was living in Southampton, at almost the opposite end of the country from my current home in Scotland. We had there a very dilapidated, ivy-encrusted structure that might once have been described as a gazebo, in which I placed the row of nests on a bench seat. This I called my 'bumblebee loft'. When I first opened the doors to the newly installed bumblebee nests, the bees poured out, eager to explore their new environment. The air filled with hovering, circling bees, but within a few minutes they had all disappeared off into the surrounding gardens where they happily began foraging, returning a little later with balls of pollen on their legs and full honey stomachs. Once the bees were experienced, which took some only an hour or two, their behaviour on leaving the nest became obviously different; instead of indecisively hovering about or circling the nest, they whizzed out purposefully and disappeared at high speed.

When the bees had had a week to settle down, I began my homing trials. I caught them as they left their nest to forage and glued a tiny coloured and numbered disc to each of their backs so that I could recognise them again. I then placed them individually in small cylindrical cardboard pots (still known to entomologists as pillboxes, from their original use) and drove them in my car to a random location. At the time I had a rather silly two-seater Toyota MR2 sports car and the stack of cardboard pillboxes would sit on the seat next to me as I roared through the Hampshire countryside. At a random location, predetermined by blindly sticking a pin in a map, I would stop by the roadside and release a batch of ten bees, noting down their numbers. Invariably and understandably they looked rather confused, and would usually circle about in much the same way as they had when they first left their nest in my bumblebee loft. Some would head straight to the nearest flowers for a quick drink after the hot journey in the car. Within a few minutes, all had disappeared into the distance. At this point I would jump back into the car and race for home. Once back at the loft I would sit and wait for the bees to return, eagerly looking at each incoming bee to see if it was one of those I had released.

Some I released within 1 kilometre of my house, and more often than not these bees would beat me back to the nest. Even from 2 or 3 kilometres the bees would often be back within a few minutes, while I got caught up in the usual Southampton traffic (bees don't have to worry about such things). From further afield, it took rather longer. I released bees up to 15 kilometres from my house and sadly none of these ever returned. I like to think that they had a nice life, freed from the burden of work in the nest, and enjoyed themselves. Perhaps they found a different nest of their species and tried to lay some eggs in it (something that worker bees sometimes do). I am probably kidding myself. Whatever happened to them, I never saw them again.

The record distance over which a bee successfully returned to the nest was 10 kilometres. I was very proud of Blue 36. It took her two days to get home. Between 3 and 10 kilometres the number of bees that made it back steadily declined, and some would take up to three days to return to their nest. White 15 was released next to a nice patch of borage flowers in the grounds of Chilworth Manor, about 3 kilometres from my house. Borage produces particularly large amounts of sweet nectar, which both honeybees and bumblebees love (the flowers also make a colourful addition to salads, and the leaves are considered by some to be an effective cure for premenstrual tension, should you need any other reason to grow some). I happened to be doing some other experiments with bees on this patch of flowers, and was returning to them at regular intervals when not transporting bees around Hampshire or sitting by the loft waiting for them to return. White 15 was one of the clever bees that made her way home successfully and swiftly, but to my amazement on subsequent days she went back to the borage flowers. Every evening she was in her nest in my garden, but she would spend her days repeating the 6-kilometre round trip to Chilworth over and over again to collect nectar.

What can we learn from all this? Firstly, that bumblebees have pretty amazing navigational abilities. Scale the feat up in proportion to size, and Blue 36's feat is equivalent to a man being taken about 1,600,000 kilometres from home and managing to find his way back again under his own power. This is more than four times the distance to the moon. White 15's repeated journey to Chilworth to collect food is the equivalent of a man circumnavigating the globe ten times just to get to the shops – and then having to come back again – several times a day. Perhaps these comparisons are a bit silly, but it is hard not to be impressed.

So how do bumblebees find their way home and, more generally, how do they navigate? This is hard to study – after all, they are small and move fast, so it isn't possible to watch. Some clues may be gleaned from observing how a worker bee behaves when she first leaves her nest – the behaviour is very distinctive. She usually flies out just 20 or 30 centimetres, and then turns to face back towards the nest entrance. She hovers from side to side, and sometimes flies a few small loops around the nest, not going more than 2 or 3 metres away. She is probably memorising the entrance to her nest, fixing the relative location of any obvious landmarks in her tiny brain (sticks, tussocks of grass or whatever). If she flies out and cannot find her way back in, then she can never bring food to the nest – and given that her role in life is to help her mother rear more of her siblings, this would be a disaster. After a few moments, she ventures off and is lost from sight, but she usually returns soon afterwards, and repeats this several times before disappearing off for longer periods. After a little while she then settles down into foraging, flying purposefully from the nest, and reappearing at regular intervals with a full honey stomach, pollen in her pollen baskets, or both.

My notion that the bees memorise the landmarks around the nest is not just idle conjecture. If you find a bumblebee nest in your garden, try putting a novel object nearby – anything will do, a tennis ball, a plastic bucket, a garden gnome. This will cause immediate but brief consternation. Instead of flying straight into the nest entrance, returning foragers will pause, fly small loops, and hover for a few moments, just as they do when they first leave the nest. If it is a large nest a swarm of them may quickly congregate. You can almost hear the cogs in their brains turning: 'Hmmm, what's going on here? I've not seen that before . . .' They usually land somewhere near the entrance and explore the last few centimetres on foot. Once they get very close, they are able to smell that they are in the correct place and will enter the nest. On future foraging trips they will then pay no heed whatsoever to the object – that is until you remove it again, at which point they will once more rememorise their nest entrance.

This behaviour makes perfect sense. The sudden appearance of a garden gnome may not be a regular occurrence in the wild, but landmarks do occasionally change, and if bees were unable to cope with this then they would be in trouble. I was once recording the traffic from a buff-tailed bumblebee nest which was down a hole near the edge of a meadow when an inquisitive cow came to investigate. I had climbed halfway into the adjacent hedge so as not to disturb the bees, but the cow was less obliging and stood more or less on top of the nest. Within a minute or two the unconcerned cow was surrounded by confused bees, both incoming foragers with pollen on their legs, unable to locate their nest entrance, and outgoing bees busily circling to accommodate this new landmark in their memories. They didn't sting her, and were probably entirely aware that she was an animate object – within a minute or two the bees had readjusted and got back to work. Of course the confusion briefly resurfaced when the cow ambled away, oblivious to the effect she was having.

When a bee is released from a box in a strange location, she behaves in much the same way that she did when she first left her nest. She hovers, performs small loops, and repeatedly returns to the release site. She is clearly confused, and understandably so. After a few moments, she starts making longer loops, disappearing from sight, and soon she is not seen again. What is much harder to do is to work out what happens between her disappearing from sight, and her arrival back at the nest (assuming she makes it).

There is something to be learned from the time it took my bees to return. As I said, from up to 3 kilometres some bees returned very quickly, in just a few minutes, but other bees took much longer, and most of the bees taken more than 3 kilometres away took days, or didn't return at all. They clearly do not have the pigeon's inbuilt GPS system. Even from 10 kilometres it would take a bee less than an hour to fly home if she knew the way. This, and the fact that many bees don't get home at all from lesser distances, suggests that something less precise is going on.

To understand what that is, and to study bee navigation in detail, we would ideally follow their tracks. This is easy enough to do for larger creatures such as birds or mammals, by attaching a radio transmitter which allows their every move to be followed. Even with the wonders of modern technology, however, it is not quite possible to obtain a radio transmitter small enough to attach to a worker bumblebee (at least not if you seriously expect it to take off and behave in anything like a normal way). The smallest transmitters are roughly the weight of the largest bumblebees. To overcome this problem, Juliet Osborne and Joe Riley, scientists at Rothamsted Research in Harpenden, have developed a system using harmonic radar to track bees and other insects. The weight of a traditional radar transmitter is largely determined by the battery. Harmonic radar doesn't use a battery at all, but instead relies upon attaching a transponder aerial to the bee. The aerial is very light, weighing in at about 12 milligrams, which is roughly 6 per cent of the bee's weight. Given that bumblebees in flight can carry up to 90 per cent of their own body weight in food, the aerial shouldn't be a major problem. That said, it is roughly 3 centimetres long (considerably longer than the bee itself), and has to be mounted vertically on the bee's back, so it looks rather cumbersome. Halfway along the aerial is a tiny electronic gadget. My understanding of the physics is rudimentary, but essentially it works as follows: the aerial absorbs any incoming radar signal, and uses the energy in this signal to generate a new signal at a slightly different frequency. A large vehicle-mounted transmitter unit with a revolving radar dish on the roof then sends out a signal which the transponder picks up and bounces back. Inside the vehicle, a bewildering and intimidating array of electronic gadgetry controls the dish and monitors any returning radar signal. The end result is a luminous green dot on a screen – the bee – which is tracked as it flies.

This system is not perfect. The aerial has to be fitted to the bee when it leaves the nest and be taken off before the bee can get back in. The aerial is bound to make it difficult for the bee to visit flowers. Also, harmonic radar only works on direct line of sight so if the bee goes behind a hedge or tree then it disappears from the screen. Moreover, the system seems to work only up to about 1 kilometre, and it costs millions of pounds to build. Nonetheless, this is still a very cool and exciting way to watch how bees explore, and Juliet's team were able to track the exact movements of foraging bumblebees on their experimental farm. Novice bees that had never previously left the nest had very characteristic flight patterns. To start with they flew erratically and stayed close to the nest, just as I had observed in my garden. They then started short exploratory flights, looping out 30 or 40 metres from the nest before returning. Each loop would go in a different direction, so that when the flight paths were plotted on a map they made a pattern very like that of a flower, with the nest at the centre and the loops resembling a circle of petals.

After exploring their local area, the bees would then range further afield, flying hundreds of metres, and stopping to explore patches of flowers. At this point many bees started to exceed the limits of detectability and disappeared off the radar screen's edge altogether. The looping flights became replaced with fast, linear flights to particular patches of flowers, which would be repeated over and over again by individual bees. They had moved from exploring the landscape to foraging. The radar work showed that foragers travelled at about 25 km/h (even with the aerial) and could maintain their linear flight paths between the nest and flower patches even in strong crosswinds, which the bees handle by flying at an angle to their intended direction; most impressive. On one occasion a bee fitted with an aerial was spotted on the radar screen but seemed to be giving a very faint signal. When it arrived back at the nest it was found to have a foxglove flower impaled on the aerial; it had obviously tried to visit the flower and the tubular petals had become stuck and come away with the bee, giving it the appearance of wearing a giant purple hat.

From this work, it was clear that bumblebee workers regularly travel more than 1 kilometre to find and bring back food. The question then is just how far? The answer is crucial to conserving bumblebees, for the distance the workers can travel for food determines whether a particular nest will thrive or die, depending on where it is relative to patches of flowers. Too far away, and they are doomed. We might plant patches of flowers in the landscape to provide food for bees, but how many patches do we need, and how far apart should they be? The answers are determined in part by the distance over which bees can forage.

Many scientists have tackled this question. The most obvious approach is somehow to mark the bees in a nest, such as with a dab of paint, and then look to see where the marked bees are foraging. This has been tried a number of times in North America and Europe, but no matter how many bees they have marked, and how hard they have then looked for them on flowers, the scientists have invariably seen very few. The vast majority seem to disappear, reappearing every little while with food. The essential problem is that the landscape is big, and bees are small. Suppose the limit of foraging range is 1 kilometre, and that a nest sends out fifty workers to gather food in this area. A circle of radius 1 kilometre spans 3.1 square kilometres; this equates to an average of one marked bee per 6.2 hectares. At this kind of density, it is perhaps not surprising that these bees are hard to find.

In addition to her radar work, Juliet Osborne tried a much lower-tech approach with her colleague Andrew Martin. This involved planting a whole field with borage, thereby creating a huge food magnet, and then placing buff-tailed bumblebee nests at different distances away. Andrew (who bears a striking resemblance to Baldrick from *Blackadder*) came up with the cunning plan of mass-marking the bees as they left their nests by making them crawl through a box filled with fluorescent powder (using a different colour for each nest). Andrew and Juliet then watched the bees in the borage field to see which, if any, coloured powder they had on their fur. They also caught bees as they arrived back at their nests, stole the pollen in their baskets, and then analysed it to see which bees were carrying borage pollen.* The experimental nests were up to 1.5 kilometres from the borage field, but even at this distance foragers were regularly seen on it and 17 per cent of the pollen arriving back at the nest was borage pollen.

This study was part of a larger, government-funded project in which I was also involved, attempting to improve our understanding of how bumblebees survive in arable landscapes. So while Juliet and Andrew were doing their borage experiment, my postdoc Mairi Knight and I were trying a different approach on the same Rothamsted farm. Our idea was to use DNA fingerprinting to look at the spread of sister bees. Mairi, a Scottish lass who had previously worked for a number of years on cichlid fish in Lake Malawi, is a whizz in a genetics lab – while I am absolutely hopeless. Although she knew nothing about bumblebees when she started working with me, she was perfect for this project since bee

* The pollen grains of many plants have distinctive and beautifully sculptured symmetrical patterns when viewed down a microscope. Borage pollen grains resemble oval pillows with a series of parallel grooves running from end to end.

DNA is much like fish DNA once it is in a tube. The basic principle of DNA fingerprinting is very simple. Sisters are closely related, and by examining their DNA it is fairly easy to identify which worker bees are sisters. Hence we set out to catch workers of a range of common bee species along a transect across the farm. Each bee was fingerprinted, and her sisters identified. We were then able to look at how far apart sister bees were along the transect, without ever knowing where the nest was that they were coming from. For example if we found two sisters 1 kilometre apart, we could say that the foraging range for this species was at least 0.5 kilometre (if the nest was exactly halfway between the two places where we caught them), but probably more. This approach suggested that some bumblebee species seem to fly much further than others; we estimated that common carder bumblebee workers flew up to 450 metres from their nests but that buff-tailed bumblebees flew at least 750 metres (and we already knew that the buff-tails could go further still if the mood took them). If bumblebee species do differ in how far they can fly to gather food, this could mean that some species are more susceptible to a lack of flowers in the countryside than others; those that cannot fly far will need patches of flowers close to their nest if they are to survive, whereas the hardy and adventurous buff-tailed bumblebee is clearly able to cover long distances in its search for food.

So to return, finally, to my homing experiments, how did Blue 36 make it back from 10 kilometres away? Our best guess is as follows. We think that in their exploratory flights, bees learn the relative positions of landmarks, just as they memorise objects close to the entrance to their nest, mentally logging the locations of tall trees, buildings, fence posts and so on. Researchers have shown that honeybees navigate better when there are obvious landmarks, using the sun as a compass to tell them which direction their nest is in relation to them. Ants are much easier to follow than bees, and ant researchers have found that displaced ants walk in long loops from their place of release (reminiscent of the paths of novice bees when they first leave their nest). It seems likely that when I released bumblebees in strange locations they did the same thing; they looped out from their place of release in search of familiar landmarks. If they were lucky enough to find one, they would quickly be able to get home. Buff-tails obviously forage quite a way from their nests, so these bees probably had a mental map of the landmarks for a kilometre or two around my house. Those bees released many kilometres from home would have had a relatively low chance of ever stumbling upon a familiar landmark; hence many never made it back, and those that did took days to do so.

Given the huge amount of energy that a bumblebee needs to fly, their ability to navigate accurately and swiftly between flower patches and their nest is the key to their survival. They are phenomenally efficient at finding their way back to rewarding flower patches over and over again. In this way the more robust species such as buff-tails can eke out a living in inhospitable, intensively farmed areas where flowers are few and far between. The key to helping our rarer species to thrive is probably simply to add more flower patches to the landscape, making it a little easier for them to find food and keep their nests well provisioned.

CHAPTER SIX

Comfrey and Smelly Feet

Take time to smell the roses and eventually you'll inhale a bee.

Anon.

In the summer of 1996 I found myself watching bumblebees, sitting amidst a dense patch of nettles and comfrey in the Itchen Valley Country Park, on the northern edge of Southampton. I had managed to get myself appointed as a lecturer in the biology department at Southampton University, and was enjoying for the first time the luxury of being paid to do research on anything I fancied.

The job of university lecturer is a pretty odd one. For a start, an ability to lecture is fairly low on the list of attributes for which the university appointing panels look. The main criterion that is used is the research record of the candidate. This explains why some university lecturers are entirely lacking in even the most basic communication skills, or indeed social skills of any kind. When I was an undergraduate I had one lecturer who, in a manner slightly reminiscent of but much less engaging than that of my old biology teacher, Mr Newton, would deliver an entire one-hour monologue with a pipe firmly clenched between his teeth and his back to the audience so that almost nothing he said was audible. Another appeared to suffer from narcolepsy and would fall asleep mid-lecture, his eyes closing and his head nodding forwards. In the embarrassingly long silences that ensued we would fidget and eye up the door, considering making a break for it, but usually after a little while he would splutter back into life.

To be fair, I also had some fantastic lecturers, but my point is that teaching skills had clearly not been high up the agenda when my lecturers were first interviewed. The situation is much the same today, and so it was in 1994 when I was interviewed at Southampton. I was a spectacularly shy child who had gone to extraordinary lengths when at school to avoid ever having to stand up in front of the class. As an undergraduate I had faked a range of life-threatening illnesses and thereby managed to avoid ever giving any kind of presentation. I was thus hardly ideal material for a job that routinely involved standing up in front of 200 or more students for an hour at a time, but that did not matter.

My interview took place on a lovely sunny day and Southampton somehow managed to impress me. It is not a beautiful city by any stretch of the imagination; in old photographs the city centre looks splendid, crammed with timber-framed pubs and the offices of wealthy shipping merchants, but some drastic remodelling in the 1940s courtesy of the Luftwaffe removed much of its charm. (Hitler again – who would have thought that the extinction of short-haired bumblebees and the demise of Southampton's ancient city centre could have a common cause.) Nevertheless, the area around the university is green and pleasant, with a very large expanse of common land covered in ponds, woods and grassy meadows, right in the heart of the city. That day, I gave a nervous and very shaky presentation to the staff of the biology department on the research I had done in an earlier postdoc position at Oxford (on the peculiar mating behaviour of death-watch beetles), and I was then grilled on my research plans. Thankfully I can remember none of the details, but the interview did not go well. I did not get the job – I was ranked fifth of the seven people interviewed. I returned to my postdoc in Oxford where I was working on how to kill caterpillars with viruses, a most depressing project. I felt dejected and rejected.

Then, four months later, out of the blue, I received a telephone call from Southampton offering me the job. The four preferred candidates had all either turned down the job, died in mysterious accidents (I have alibis) or were otherwise no longer available, and suddenly I was in the frame. I was delighted. Naively, I failed to negotiate at all on salary, accepting the job before any figure had even been mentioned. I started as a lecturer in January 1995 and I was to spend the next eleven years working there.

Although Southampton may be few people's idea of an attractive place, Hampshire is a lovely (if rather crowded) county. Just to the west of Southampton is the New Forest, with its mix of ancient woodlands, windswept heaths and quaint chocolate-box villages; a fantastic place for a lover of wildlife. It is a particularly good hunting ground for bush-crickets, for it harbours a number of rare species. On my weekends I spent many hours trying to track down the Roesel's bush-cricket, a beautiful emerald-green-and-black beast which inhabits the salt marshes on the southern edge of the forest. The males make a high-pitched hum which is almost impossible to pin down, at least to a human, but is presumably effective at attracting females of the species. I became interested in coneheads, slender bush-crickets with rather angular heads (and a great name). There are two species in the UK, short-winged coneheads and long-winged coneheads, but the short-winged ones can sometimes have long wings, just to confuse matters. For unknown reasons both species seemed to be doing very well in the mid 1990s, using their long wings (when they had them) to spread northwards from the south coast. If I had another lifetime to live I might spend it studying crickets.

Just north from Southampton are the South Downs, the chalky

hills that run from Hampshire to Kent. Although much of the land is now intensively farmed, there are still fantastic pockets of flower-rich chalk grassland, relics of a time when the Downs were all one vast flowery meadow. One of my favourites remains Yew Tree Hill, a little nature reserve owned by Butterfly Conservation which is awash with wild flowers through the spring and summer, including huge numbers of the beautiful purple spires of pyramidal orchids. I performed many studies of bumblebees there while I was at Southampton. The Downs must once have been the most splendid place.

Running south from there to the sea are many crystal-clear streams, some of them used heavily for rearing watercress, a local speciality which needs unpolluted waters. Several of these streams converge to form the Itchen, which flows down past affluent Winchester into the Solent near Southampton. It is a lovely, clear river, with long green streamers of water crowfoot which in summer end in delicate white blooms just above the rippling surface. Trout and elegant grayling (my favourite fish) sport among the waterweeds, eyeing up the clouds of bright blue damselflies that skim above the water. Half a mile or so before the clear waters of the Itchen merge into the murky tidal soup that is the Solent, they flow through the Itchen Valley Country Park, where I found myself sitting in the summer of 1996, watching bees.

The Itchen Valley Country Park is an area of lowland water meadows, ditches, streams and woodland, hedged in by roads and housing estates. It was not a peaceful spot then and is less so now; there is a constant drone of traffic from the nearby M27 motorway, and planes roar low overhead every few minutes, taking off from the nearby Southampton airport. Nonetheless it has plentiful wildlife, with otters, water voles, and swarms of the gorgeous demoiselle dragonflies along the riverbank. I had noticed that comfrey flowers, both there and elsewhere, are very variable in colour; some are white, others pale mauve, dark purple and just occasionally bright pink. This is quite unusual for a wild flower – most tend to be uniform in colour, more or less. I was curious to find out whether different insects preferred particular colours, and so I had sat down to watch. The visitors to comfrey are almost all bumblebees. Comfrey flowers are quite deep, so that only long-tongued bumblebees such as the garden bumblebee and the common carder can reach the nectar, but I noticed that white-tailed and buff-tailed bumblebees, which are short-tongued, would readily bite a hole in the side of the flower and steal the nectar (something known as nectar robbing). If I listened closely I could actually hear their jaws chomping through the petal. Early bumblebees would also visit the flowers; they are short-tongued but did not bite holes in the flowers. Instead they rely on those already created by the buff- and white-tailed bees. Almost every comfrey flower had a neat hole in the side. All of this was bad news for the flower, for robbing bees do not come into contact with the reproductive parts of the plant and therefore do not pollinate it. Fascinated, I watched for many hours, recording the sequence of visits by each bee to the pink, purple or white flowers. It turned out that none of the bees seemed to care what colour the flowers were, readily flitting from one to another, showing no preferences; not the most exciting result.

I was about to give up on all this when it dawned on me that I had been seeing something else that was very odd, although it had nothing to do with the colour of the flowers. Dozens of bees were buzzing around in the comfrey patches, flitting swiftly from flower to flower. Very often they would fly up to a flower, hover in front of it, but then fly away without landing. They might do this to three or four flowers before finding one that was apparently to their taste; then they would land and feed. What were they doing? I was puzzled. To try to find out, I measured the nectar in these rejected flowers. This is a fiddly business. The usual approach is to use a very narrow glass tube called a capillary tube, about 1 millimetre in diameter. This has to be carefully pushed into the flower in exactly the way in which a bee inserts its tongue, so that the tip of the tube touches the tiny drop of nectar in the nectary at the bottom of the flower. When this happens the nectar is sucked into the tube all on its own, due to something known as capillary action. By measuring how full the tube is with a ruler, one measures the amount of nectar. So I watched bees, and measured the nectar in hundreds of the flowers that they rejected. I did the same with the flowers that they landed on, quickly shooing the bee away after it landed but before it could drink. It turned out that those flowers that were being rejected had less nectar in them than the flowers the bees landed on. Somehow they could tell which flowers had the most rewards, and were avoiding landing on those that didn't have much in them. How were they doing it? I was intrigued.

Luckily, at about this time I managed to get funding from the university to support my first PhD student. By coincidence, Jane Stout, a recent graduate with a first-class degree in Environmental Sciences from Southampton, had just returned from an expedition in Tanzania and was looking for a job, so I persuaded her to have a bash at doing a PhD. Jane and I were to spend much of the next four years trying to work out exactly what these bees were doing.

Bees that are collecting pollen can often see which flowers have most reward, because in many flowers the anthers that produce the pollen are visible from a distance. Pollen is usually brightly coloured: yellow, orange, white or purple. Look at a rose, geranium or bramble, and to the human eye it is obvious that some flowers have more pollen than others (the ones with little pollen usually being ones which have recently been stripped by a bee). Bees have pretty sharp eyesight, and we found that pollen-collecting bees do indeed quickly learn to spot the most rewarding flowers. But the nectar in comfrey is hidden in the bottom of the flower (or, rather, in the top of the flower, since the flower hangs down like a bell), and in any case is a colourless liquid. How could bees tell which flowers had most nectar in them? In fact, it turned out that they couldn't. We took flowers that had just been emptied by a visiting bee, and added more nectar to them (having carefully sucked it out of another flower). These flowers were still rejected by passing bees. We also covered flowers with netting for several hours to keep bees away and allow them to fill with nectar, and then artificially sucked the nectar out with capillary tubes. Bees readily landed on these empty flowers, only to be disappointed on finding nothing inside. So it seemed that the bees weren't able to tell how much nectar was in flowers, yet in natural situations without scientists messing everything up for them they somehow only landed on the full ones. How on earth were they doing it?

After a lot of pondering, I realised that the bees must somehow be able to tell which flowers had recently been visited by another bee. Regardless of whether it was full or empty, bees would not land upon a flower recently visited by another. Similarly, even if they were empty, bees readily landed on flowers that had been covered in netting to keep them away. So it had to be something to do with the bees visiting the flowers. But what? Bees didn't leave behind any visible mark when they visited a flower, unless they bit a robbing hole, and almost every flower already had one of those so that couldn't be the clue. Perhaps they could smell where other bees had previously been? We tried washing the feet of each bee with a tiny drop of solvent, then pipetting the drops on to flowers – and hey presto, those flowers were rejected by other bees. If we pipetted only solvent on to the flowers, there was no effect. So it turns out that bees leave behind a smelly footprint when they land, which any subsequent passers-by can detect, alerting them to the fact that the flower has recently been emptied.

So why do bees have such smelly feet? In fact all insects are covered in an oily liquid that helps to keep them waterproof. It is not just their feet; their entire bodies are bathed in a thin layer of oil. Each insect species has its own particular blend of these oily hydrocarbons, leaving tiny traces on anything they touch. The antennae of insects are finely tuned to detecting these traces, so that they can readily smell just a few molecules in the air around a flower, and this warns them that it is not worth the bother. By saving time that would be wasted climbing inside empty flowers, bumblebees can gather more nectar per hour and their nest can grow more quickly.

Of course flowers refill with nectar and using smelly footprints to detect empty flowers would only work if the footprint wore off. This seemed to be exactly what happens. We recorded how long a comfrey flower remains repellent to passing bees after it has been visited, and found that the effect seems to wear off after about forty minutes. We then carefully measured how quickly the flowers refilled with nectar, and found that they took between forty minutes and an hour to refill. In other words there was a pretty good match between when bees would visit a flower and when it was likely to be full, or nearly so.

With our feeble sense of smell, this all seems terribly impressive. However, there is more. It turns out that different flower species refill with nectar at different rates. For example, borage produces nectar very fast, whereas comfrey is middling, and bird's-foot trefoil is very slow. When feeding on borage, bees start revisiting a flower just two minutes after the previous visit, and again this roughly corresponds with the time it takes to refill. On comfrey, as I mentioned, bees revisit flowers after about forty minutes, while on bird's-foot trefoil a flower seems to remain repellent for at least 24 hours.* Yet the footprints are the same.

How does this work? It seems that bumblebees are able to tell how old a footprint is, perhaps by the strength of the smell, and that they learn an appropriate threshold for any particular flower. This is helped by the fact that individual bees tend to specialise for several days at a time, and sometimes for their entire life, in visiting just one flower species over and over again, gaining a lot of experience. So a bee visiting borage quickly learns to ignore all but the freshest smelly footprint, while a bee visiting bird's-foot trefoil learns that a flower that smells even faintly of another bee's feet is best avoided.

We later looked at other insects, and found that this system seems to work across species. Both honeybees and all of the bumblebee species that we studied seem to use footprint smells to judge which flowers to visit, and are able to recognise the footprints left by other species just as well as their own. This of course makes sense, for it does not matter who has emptied a flower – it is still empty. Bees even seem able to tell when a flower has been visited by a hoverfly.

It is probable that without their ability to detect and avoid empty flowers, bumblebees would struggle to survive. In our comfrey patches in the Itchen Valley Country Park there were hundreds of bumblebees foraging together, and most of the flowers were empty or nearly so at any one time. Landing, pushing her tongue into the flower and then taking off again all takes time and energy for a hungry bee. Even saving a fraction of a second can, cumulatively, pay huge dividends, for each bee must find tens

* Bird's-foot trefoil is so named because the seed pods look remarkably like the three-toed foot of a bird, not because they smell of birds' feet!

of thousands of full flowers per day if she is going to fuel all of her flight and bring back a net return of nectar for the nest.

This aspect of bee behaviour was tremendous fun to explore over a number of years, but it turned out that we got one part of it wrong. Jane and I stopped working on bees' smelly footprints in about 2000, conceitedly thinking that we had pretty much wrapped up most of the interesting angles on this. I started working more on the ecology of rare bees, and Jane got herself a lectureship at Trinity College Dublin, where she has since made a name for herself studying the pollination of invasive weeds such as rhododendron. In 2005 I found myself at a conference in St Petersburg chatting to Thomas Eltz from Dusseldorf University. He had been analysing the speed at which the chemicals in bee footprints evaporated from flowers, and had found that it was far too slow to fit with our explanation. The compounds are very large and not very volatile at all, so that they remain on the flower more or less indefinitely. In fact he found that flowers accumulate chemicals from successive bee visits so that, with careful analysis, the flower can provide a record of all the insects that have visited it during its life.

This then begs the question: how can bees tell how old a scent mark is? If it doesn't evaporate, then it will not fade over time, and it should be impossible to know whether the flower has had time to refill with nectar, yet clearly the bees somehow do know. We still don't have a definitive answer. My guess is that perhaps the chemicals deposited by bees' feet slowly sink into the waxy coating of the flower – for just as bees are waterproofed with oils, so flowers are waterproofed with oils and waxes – and by sinking into the flower's surface they become less detectable, and effectively fade even though they are still present.

Interestingly, the same smelly footprints, when encountered in a different context, have a different meaning. Bee footprints around the entrance to the nest, rather than having a repellent effect, help returning foragers to find their way inside. It seems that bumblebees learn to interpret a particular smell in an appropriate way, depending on the circumstances in which they encounter it.

It is humbling to reflect that though a bumblebee has a brain smaller than a grain of rice, it has powers of perception and learning that often put us mammals to shame. Next time you are sitting in your garden while the bees are visiting your bee-friendly plants (if you haven't got any, I hope you'll plant some next spring), take the time to watch what they are doing. You will quickly notice them dismissing some flowers after a quick sniff. But I'm sure that there is still much more to learn. Certain flowers seem to be visited by particular species of bee, and often we have no idea why. Individual bees may collect pollen or nectar or both, and seem particularly disposed to collect one or other depending on the flower, but again we often have no explanation for their choices. On some days, even in high summer when the weather is fine, foraging bees may suddenly become scarce, as if they have all decided to go on strike for a few hours; we do not know why. Bumblebees are one of the most familiar and intensively studied of all the insects on earth, but there is still an enormous amount that we do not understand about their lives.

CHAPTER SEVEN
Tasmanian Devils

Burly, dozing humblebee,
Where thou art is clime for me.
Let them sail for Porto Rique,
Far-off heats through seas to seek.
I will follow thee alone,
Thou animated torrid-zone!

Ralph Waldo Emerson
(American poet)

In Tasmania the first bumblebee was recorded in 1992. There was no mistaking her, for the local bee species are tiny – small enough to hide under a grain of rice, mostly drab and not very furry. Nor do bumblebees naturally occur anywhere near Tasmania, for they are mainly creatures of the northern hemisphere. So these new furry giants would not have escaped notice for long, especially since they first appeared in the gardens of Hobart, the most densely populated area of the island.

These new arrivals were quickly identified as buff-tails. Now buff-tailed bumblebees only actually have buff tails in the UK. In the rest of Europe, buff-tailed bumblebees have white tails, which, as you might imagine, makes them awfully hard to distinguish from white-tailed bumblebees. Anyway, these were buff-tailed buff-tails, which meant that they had to have originated from the UK, 10,500 or so miles from Tasmania. If your geography is a little rusty, Tasmania is the southernmost state of Australia, a roughly triangular island floating 150 miles off the south coast of the mainland, with New Zealand lying about 1,500 miles to the east. So how did they get there? The answer is not quite as mysterious as all that, for as I have already mentioned, English buff-tailed bumblebees have been living quite happily in the wild in New Zealand since 1885 or thereabouts. The Tasmanian bumblebees presumably came from there, but this is still a 1,500-mile journey against the prevailing wind, across a stormy and cold Tasman Sea. With the best will in the world they could not have flown.

We may never know how they made the journey. They may have been accidentally transported on a ship; a young queen may have hibernated in a plant pot and been brought over with some nursery plants. But I find this unlikely. It is probably no coincidence that in about 1988 tomato growers the world over turned to using bumblebees to pollinate their crops. Researchers in Belgium and Holland had discovered that bumblebees are fantastically efficient at pollinating greenhouse tomatoes, and they had also worked out how to breed buff-tailed bumblebees in large numbers. As a result, bumblebee-rearing factories quickly sprang up in Europe and then in North America and Asia to cater for the demand. The only tomato growers left out of the bumblebee bonanza were those in Australia, where there are no native bumblebees, and where importing foreign species is strictly forbidden. Tomato growers on mainland Australia still have to hand-pollinate their plants; teams of workers are employed, each of whom is equipped with a slender vibrating wand. Every flower has to be touched with the tip of the wand if it is to set fruit. As might be imagined, this is tedious work in a large glasshouse – some commercial operations cover hundreds of acres and contain literally millions of tomato flowers – and the labour costs are substantial. Bumblebees are not only much, much cheaper, but the tomatoes produced by bumblebee pollination are also larger and apparently taste better than those pollinated by humans, so in the late 1980s, Australian tomato growers suddenly found themselves at a distinct disadvantage. That bumblebees mysteriously arrived in Hobart in 1992 may just be coincidence. Bumblebees had failed to cross from New Zealand to Tasmania for 100 or so years, but suddenly they managed to do so just after it was discovered that they had huge commercial value. Draw your own conclusions.

Yet why shouldn't bumblebees be imported to Tasmania and, for that matter, to mainland Australia? After all, they are cute, furry and beneficial insects which pollinate crops and wild flowers. Who wouldn't want them? The answer is that man has a rather poor record with regard to introducing non-native species, and nowhere more so than in the Antipodes.

Australia is a remote island, with a remarkable and unique fauna, of which the bizarre marsupial mammals are the best known. Only distantly related to the mammals found in most of the rest of the world, they evolved into many wonderful forms: kangaroos, bandicoots, koalas and numbats, among numerous others (and top marks to the Aboriginal people for giving them such memorable names). Australia also has thousands of indigenous bees, butterflies, flowers and so on. New Zealand is the same, only more so as it is even more isolated than Australia, so no mammals got there apart from bats (which, of course, can fly). There, giant flightless birds took the roles of large mammals, and enormous crickets (known as weta) evolved to fill the role of mice. When Captain Cook first arrived, New Zealand was clothed in verdant forests teeming with birds and insects, almost every one of which he had never seen before; they were all unique to New Zealand.

Tragically, the early European settlers in Australia and New Zealand soon became homesick, pining for such familiar creatures as foxes, rabbits and hedgehogs. They formed Acclimatisation

Societies, dedicated to introducing as many non-native plants and animals as possible, and awarding medals to those who were most successful or dedicated in their efforts. With spectacular naivety, in New Zealand they even experimented with introducing zebra and giraffe. Aside from these, the introductions were remarkably successful. When I first flew to Christchurch (with Mick Hanley in 2003) it all looked tremendously familiar. We had left behind a cold and drizzly winter's day in London and arrived to glorious summer sunshine, but there the dissimilarity ended. Perhaps not surprisingly, the buildings in Christchurch look decidedly British. After all, all the older ones were built by British immigrants. The trees lining the streets are limes and planes, just like those in London. The birds chirping from their branches were blackbirds, sparrows, thrushes, starlings and greenfinches. We hired a car and drove out of the city across the Canterbury Plain, the main agricultural region of New Zealand. So far as I could tell as we rocketed along with Mick at the wheel, the blurred trees, roadside flowers and farm animals that we shot past appeared to be much the same as those at home. When we stopped – in my case, saying a silent prayer of thanks that I had survived the journey so far – a pair of goldfinches flew by, and skylarks trilled overhead. Had it not been for the snowy mountain peaks in the distance, we could have been in Cambridgeshire or any other flat, agricultural part of lowland England.

It was several hours and 300 miles before we saw something new; a small friendly bird which my bird guidebook suggested was a fantail. New Zealand is famous for its beautiful scenery (think *Lord of the Rings*), but it is an ecological holocaust. In 200 or so years we have wiped almost all the native animals from vast tracts of the country, clearing native forests to make way for farming, and especially for sheep ranching. In the remote mountains there are still forests of native trees, but the birds that once inhabited them have mostly gone, eaten by the introduced foxes and stoats, against which the native species seem to have no defence. Famously the kakapo, a chubby moss-coloured flightless parrot, attempts to escape from predators by shinning slowly up the nearest shrub and then jumping from the top, only to thud down to the ground a yard or so from where it started. Hardly enough to throw a wily fox off its trail. Hence this once widespread New Zealand bird was down to about sixty individuals at the last count.

The situation is almost as bad in Australia, where the native marsupials have struggled to cope with introduced pigs, rabbits, foxes, cats, camels, dogs and goats. Vast tracts of forest have also been cleared to make pasture for cattle and sheep, and have subsequently become overrun with European weeds such as the appropriately named Patterson's curse,* while the tropical north-east is swamped with the South American shrub *Lantana*. Perhaps most famously, cane toads (also from South America) have bred in their countless millions in Queensland, from where they are spreading remorselessly southwards, consuming everything that they can fit into their capacious mouths.

You get the picture: introducing non-native species can have disastrous consequences. But surely bumblebees are beneficial insects? What harm could they do? Well, probably quite a lot actually, although we don't know for sure. That was why I went to Tasmania in January 1999, with my PhD student Jane Stout. The introduction of bumblebees there was a huge experiment

* Much loved by honeybees but unpalatable to grazing animals, Patterson's curse was introduced to Australia in the 1880s by Jane Patterson, an early settler. She innocently brought the seeds from Europe so that she could grow the pretty flowers in her garden, but the plants rapidly spread into the surrounding pasture. The latest estimates suggest that this one weed now costs Australian farmers $30 million per year.

(albeit a very poorly designed one, with only one replicate – Tasmania – and no control). Here was an opportunity to see at first hand what an introduced species might do (and, of course, to escape from the British winter for a few weeks).

We first needed to find out exactly how far the bees had spread. At that time they had been on the island for about seven years, and as the anecdotal records suggested that they had been spreading steadily north and south from Hobart, we hired a tinny little car and set about driving the length and breadth of the island.

Tasmania is stunningly beautiful. The population of less than half a million mostly live in or near Hobart in the south-east, leaving much of the remaining 26,000 square miles very sparsely populated. To the west, the island becomes mountainous and largely inaccessible, clothed in temperate rainforests containing dense stands of giant tree ferns, above which tower the tallest flowering plants in the world, the mountain ash, which grow to well over 300 feet. True to their name, these forests rely on the near-incessant rain coming in on the prevailing westerly winds; they have a damp, gloriously musty scent and prehistoric feel. By contrast, the north and east of the island comprise lower-lying, rolling countryside, with much of the natural forests cleared to make way for farming, mainly sheep ranching. The coasts have some of the most spectacularly scenic sandy beaches I have ever seen, although even in summer the water is a tad chilly. The coastline is somehow made more atmospheric by the knowledge that there is nothing but icy ocean between Tasmania and the South Pole.

One of the joys of visiting Australia for the first time is seeing flocks of wild parrots. Even in the centre of Hobart, flocks of swift parrots live up to their name by rocketing like emerald-green darts from gum tree to gum tree, acrobatically drinking nectar from the flowers and squawking raucously. Beautiful red, yellow and blue eastern rosellas also stride cheekily among the picnickers in the parks, scavenging scraps of food. I had never seen a parrot outside a cage before, and it was wonderful to see these brilliantly coloured creatures living free. I was also particularly keen to see some of the island's mammals as we toured. I had read that Tasmanian devils were not uncommon, and I was desperate to see one (perhaps just because I was very fond of the cartoon series, although I didn't really expect them to spin on the spot like miniature tornadoes). Unfortunately these mammals are largely nocturnal, which makes live ones hard to spot, but in an echo of my childhood, dead ones littered the roads. Just as the fauna of Australia and New Zealand has coped poorly with introduced predators, so it seems that the Tasmanian marsupials are spectacularly inept at avoiding cars, for we saw hundreds of corpses. After a few days I gave up hope of ever seeing one alive, and took to photographing the roadkill, building up a fine collection of photographs, including squashed Tasmanian devils, pademelons, bettongs, possums, wombats and potoroos (such wonderful names).

Eventually we did come across one live mammal, in the form of the amazing echidna. From a distance echidna resemble rather chubby, rounded hedgehogs, but close up there are a number of obvious differences. The snout is much longer, tapering to a blunt tip. The spines are enormously thick, more like the quills of a porcupine. And the spade-like feet possess massive claws, enabling the echidna to rip apart termite nests or to dig vertically downwards in times of trouble. In fact once we had seen one, we suddenly starting spotting them on most days. Echidnas are monotremes, members of an obscure group of mammals which includes only themselves and the platypus, both famous for being egg-layers. We eventually met five echidnas in the course of our travels, each of which delayed us for quite some time as we felt compelled to watch them lumbering and snorting about their business like miniature spiny bulldozers. Now when I am asked to name my favourite mammal, I always answer echidna.

Of course we were supposed to be looking for bumblebees, and perhaps part of the reason that we were so easily distracted is because it was proving difficult to establish the distribution of a smallish creature such as a bumblebee in a few weeks over a huge area. Whenever we found one, it was easy enough to mark its location on the map. But when we didn't, it was hard to say whether that was because there weren't any, or because we'd been unlucky, or because there weren't many bee-friendly flowers in that particular place. However, it quickly became apparent that one of the best ways to find bumblebees was to find a well-tended garden with lots of flowers in Hobart itself. Tasmanians are enthusiastic gardeners, perhaps because the climate is particularly benign, but for whatever reason Tasmanian gardens are often rather splendid. Lavender in particular grows well and bees love it. In any garden in Hobart, so long as there were a few lavender bushes, we found that we were guaranteed to see a bumblebee within a few minutes.

So our strategy became one of finding the most colourful garden in each town or village as we drove around Tasmania. We would then either peer in over the fence or, if that wasn't possible, knock on the door and ask if we could look for bumblebees in their garden. I guess that crime rates there are low. I've tried the same approach in Britain and generally received short shrift from homeowners who clearly thought this a flimsy and highly implausible excuse for casing the joint. A strange man on the doorstep holding what is often mistaken for a very odd-looking fishing net in one hand and a cluster of urine sample tubes in the other rarely gets a good reception. In contrast, we were universally welcomed by Tasmanians – perhaps helped by the fact that I usually got Jane, who has a very friendly smile, to knock on the door, and also because experience has taught me to hide the tubes. The only downside to the Tasmanians' welcoming attitude was that we often then became embroiled in very long conversations obliging us to explain what bumblebees were and what we were doing, listen to long accounts about other interesting creatures that they had seen in their gardens, have a cup of tea and so on – all very agreeable but not terribly productive.

The more rural areas, particularly in the west of Tasmania, were harder because there we encountered huge areas with no people, and hence no gardens. We would search for bees on patches of flowers growing by the roadside, one of the more striking of which were tree lupins. These are the North American relatives of the lupins we commonly grow in our gardens, but they are much larger as one might guess from the name, although 'tree' is pushing it a bit – they are rarely more than 6 feet tall. They were apparently introduced to Tasmania in the 1920s in an attempt to stabilise the coastal dunes; in their native California, tree lupins thrive on very sandy soils. (Of course this begs the question why dunes need to be stabilised. They'd presumably been perfectly happy for thousands of years being unstable.) Anyway, the flowers are bright yellow, and stands of tree lupins make a splendid sight. As bumblebees love them, we took to searching for them in the areas where there were no gardens, and spent a very pleasant couple of weeks touring Tasmania and producing a distribution map. It was clear that the bees had spread a long way; roughly 60 miles north and south (reaching the southernmost tip of the island), and about 50 miles west. So far as we could tell, they had not reached the north of the island, and we could find none in the dense forests to the west. The bees seemed to be largely confined to places where there were either gardens or lots of European or North American weeds.

This is all well and good, but what harm might these friendly, furry beasts be doing? There are a number of possibilities. The most obvious danger was that they might out-compete the native species, for bumblebees, along with all other bees, feed only on nectar and pollen; and as they need a lot of nectar to fuel their activities, this inevitably means less for others. In Tasmania, the native nectar-feeding fauna includes hundreds of small bee species, other insects such as flies, beetles, butterflies and so on, several varieties of parrot, and also ten species of long-billed birds variously called honeyeaters, wattlebirds and spinebills (four of which occur only in Tasmania). If nectar and pollen are plentiful then adding bumblebees into the mix may not matter; but if there is a shortage, any food taken by bumblebees means less food for the locals, quite possibly with disastrous effects if the creature concerned is already suffering from the impacts of logging or introduced predators.

A second possibility is that adding bumblebees to the Tasmanian mix might reduce pollination of native plants. You might be wondering how having bumblebees could *reduce* pollination – after all, they are very good at the job. Let me give you an example. Honeyeaters specialise in feeding on nectar; they are the Australian equivalent of a hummingbird (although they do not hover), with long, down-curved bills for reaching into deep flowers. Many Australian plants have specifically adapted to such pollination, evolving deep tubular flowers in which only the honeyeaters can reach the nectar. Typically, as the bird probes down it receives a dab of pollen on its forehead, and as it travels on from flower to flower it spreads the pollen, so fertilising the flowers. Now imagine a bumblebee faced with such a flower. Its tongue is not long enough to reach the nectar, but buff-tailed bumblebees are adept at nectar robbery – by biting a hole in the back or side of the flower they can access the nectar, but go nowhere near the reproductive parts, so they do not pick up any pollen, and the flowers do not get fertilised. With lots of bumblebees at work, it is easy to imagine harmful effects on both the honeyeaters and the plants.

A third possibility is that bumblebees might improve pollination of the undesirable alien weeds with which Australia is rife. This is probably the simplest and most likely impact. One of the commonest routes for the introduction of non-native weeds was as seeds in the hay brought over on ships with livestock from Europe, the result being that almost all of the common hayfield plants of Europe now occur in Australia. Many of these weeds are naturally pollinated by bumblebees in their native range. Some of them are welcome in Australia; clover, for example, is a valued food for livestock. But many, such as thistles and gorse, are invasive weeds. Imagine then the likely consequences of introducing a more effective pollinator, one with which these plants evolved for millions of years in Europe. Plants such as gorse have flowers that have adapted to be pollinated by large furry bees. The tiny native Tasmanian bee species might well be quite hopeless at pollinating these flowers. But with bumblebees on the scene, it is possible that these weeds could produce more seeds and spread faster. One might even imagine that there might be 'sleeper weeds': non-native plants which have remained rare for decades in Australia because they rely entirely on bumblebees for pollination. Add bumblebees, and these sleeper weeds might awake, rampage and spread.

Was any of this actually happening? We needed more time if we were to get answers, and so it was that in December 1999 I returned to Tasmania with Jane and an extra pair of hands in the form of a second PhD student named Andrea Kells. This time I had managed to negotiate a sabbatical and could stay for the whole of the Antipodean summer if need be.

We had a two-step plan of action. First, we decided to see whether the native bees had become less common in places where bumblebees had arrived, by comparing their numbers at sites within the expanding range of the bumblebee with sites just outside the current range. The second stage focused back on the tree lupins. Lupins are a typical bumblebee-pollinated flower; those in my garden in

Southampton seemed to be pollinated solely by bumblebees, and it is said that tree lupins in California are also bumblebee-pollinated. We had also read that tree lupins in New Zealand and Chile are considered to be major weeds (although admittedly rather pretty ones). Tasmania has a similar climate to New Zealand, yet here tree lupins seemed to be generally rather scarce, and that led us to wonder whether the New Zealand lupins had benefited from 100 years or so of bumblebee pollination, whereas those in Tasmania were lacking their main pollinator until very recently. Could it therefore be that Tasmanian tree lupins were a 'sleeper weed', and that the bumblebees' arrival would awaken them to their true weediness potential? The answer lay with finding tree lupins both within and without the current bumblebee range, with the aim of measuring whether they were setting more seed where bumblebees now occurred.

This all required a lot of driving along the quiet and winding Tasmanian roads. Which in turn of course allowed me to indulge my pie obsession as we paused to lunch on delicious seafood pies, filled with luscious juicy scallops from the chilly coastal waters. We also saw many more squashed animals, a few live echidnas and, on one memorable occasion, a stunningly beautiful tiger snake crossing the road, its ebony scales glinting in the sunlight. These snakes are deadly, but I could not resist getting out of the car to have a good look and timidly snap a few pictures. It didn't seem to be remotely interested in me, and slithered gracefully off into the undergrowth.

To count the numbers of native bees in different places we carried out one-hour timed searches, during which we recorded every bee that we saw, catching those that we didn't recognise for identification later. We also stuck up a yellow A4-sized 'sticky trap' on a tree or telegraph post at each location. These are simply bright yellow plastic sheets covered in phenomenally sticky glue – insects mistake them for enormous rectangular flowers and when they investigate become permanently stuck. After a week we came back to collect the traps. This quickly turned out to be rather scary. Although I have spent virtually my whole life chasing bugs of one sort or another, I have to confess to a bit of an aversion to spiders. I think I picked this up from my mother, who to this day becomes near hysterical at the sight of even the most innocuous spider. Australia is well known for its poisonous snakes and spiders, and amongst the most frightening of them all in appearance is the huntsman. They are not actually poisonous, but they are massive, crab-like, hairy beasts with huge curved fangs and eight eyes glinting with malicious purpose (or at least that is how my fevered imagination remembers them). They move like lightning, and their favourite habitats are the telegraph posts and tree trunks along which they prowl at night in search of their prey. They have flattened bodies for sliding into cracks and gaps under dead bark, and they obviously liked to hide under our sticky traps. Some would also walk over our traps at night and become stuck. The very largest ones were so powerful that they could tear themselves off the glue and escape, leaving a trail of bristles and footprints. All of this made collecting our sticky traps a harrowing and adrenalin-laced business for an arachnophobe. Almost every trap had at least one huntsman stuck to it, or one ready to shoot out from underneath as soon as the trap was moved. Worst of all were the ones where hairy footprints spaced a hand's width apart showed that a particularly large spider had pulled itself off the glue, and was no doubt lurking somewhere nearby waiting for revenge.

The spiders aside, our bee-counting revealed something we hadn't anticipated. Just over half of all the bees we counted were honeybees. Honeybees are the anorexic cousins of bumblebees, smaller, slimmer and much less furry; they are usually fawn or tan in colour with vague darkish stripes. Now honeybees are also not native to Tasmania. They come from Europe and the Middle East, but they have been domesticated by man since prehistory (ancient

Egyptian hieroglyphs contain carvings of bee hives). Because they have always been highly valued both for crop pollination and for their honey, we have imported honeybees to every country in the world except Antarctica. They were introduced to Australia in 1821, and there is now a very big honey industry. Tasmanian leatherwood honey is particularly delicious, and fetches a premium price.*

When we analysed our data, which contained many thousands of records of insects at over 120 sites, we found no measurable effect of bumblebees whatsoever. Places where bumblebees had recently invaded had just as many native bees (both in terms of numbers of different species and numbers of individual bees) as places where bumblebees had not yet arrived. By contrast, the honeybees were having a major effect. Wherever we found honeybees, there were on average only one-third as many native bees. With hindsight, this result is perhaps not surprising. Honeybees were much more numerous than bumblebees, and as they also live in much larger colonies, if any non-native species was going to be competing with native bees by using up lots of pollen and nectar then honeybees were always going to be the more likely candidate. They also have quite short tongues, as do all the native bee species, so they tend to feed on the same flowers, whereas bumblebees have longer tongues and so tend to choose different flowers (although there is lots of overlap). So if bumblebees were impacting on native species, they were far more likely to be doing so on birds such as honeyeaters, who favour deeper flowers.

Discovering that honeybees seemed to be having a harmful

* There is a wonderful honey shop in the tiny village of Chudleigh in northern Tasmania. It sells over fifty local varieties, all of which are laid out for tasting, and a huge range of other bee- and honey-related products, including a baby's bee outfit which I couldn't resist buying for my youngest son, and in which he looked ridiculously cute.

effect on native species put us in a rather delicate position. Bumblebees had recently arrived in Tasmania, and had we found that they were causing harm to native bee numbers nobody would have been too upset or surprised. On the other hand, honeybees are highly valued and make a significant contribution to the economy. Many people make their livelihoods from keeping them, and they no doubt contribute to pollination of many crops. Understandably, these folk would not appreciate a bunch of Poms swanning over for a few weeks and then declaring their beloved honeybees to be undesirable aliens. But on the other hand, there is no doubt that honeybees are a non-native species. It also seems to me common sense that flowers can produce only so much nectar, and that there can therefore be only so many bees in a particular habitat. Something has to give, and that thing is likely to be the local flower-visiting insects. This is just as true elsewhere in the world as it is in Tasmania. There are huge honey industries in New Zealand, mainland Australia (I've always thought eucalyptus honey tasted rather medicinal but it is very popular), and throughout the Americas, and in all these places honeybees are non-native.

Let me bore you with a few figures. A single honeybee hive contains 50,000 workers or more, and it is common for beekeepers to put twenty hives in a single place – 1 million bees. A single honeybee nest harvests up to 60 kilograms of pollen and 150 kilograms of nectar per year. At high hive densities, honeybees can harvest up to 22,500 kilograms of honey per square kilometre. The New Zealand honey industry produces 8,000 tonnes of honey per year from 227,000 managed hives, or thereabouts. Some of these vast quantities of honey are obtained by the bees foraging on crops which are in turn pollinated, but a large proportion comes from the bees visiting wild flowers. With this much honey being taken by honeybees, it seems obvious that there is likely to be an impact on other creatures that need nectar.

Even in places where honeybees naturally occur such as the UK, beekeepers often create unnaturally high densities. A student of mine, Kate Sparrow, recently found that bumblebees in Scotland tend to be smaller in places where there are lots of honeybees, presumably because of competition for food.

I do not want to pick a fight with beekeepers. Most are very fond of bumblebees and other insects. They are invariably aware of the shortage of flowers in the countryside and keen to support efforts to make it more bee-friendly. In short, in many ways beekeepers are the natural allies of a bumblebee conservationist such as me. But there is no denying the potential conflict and I am sometimes saddened by the strong reaction of some beekeepers to the merest hint that their bees might occasionally do harm. It is the simple truth. In fact, there are rather few places where there is likely to be any major conflict between conservationists and beekeepers. In most of the farmed countryside, be it in Tasmania, New Zealand or the UK, the benefits that honeybees provide through pollinating crops and producing honey greatly outweigh any small impact they might have on other insects. But in a few special places, it might be wise not to station honeybee hives. Imagine a small nature reserve in Tasmania, for instance, supporting the last-known surviving population of the striated dongle bee (a hypothetical creature, in case you were wondering). Would this be a sensible place to put twenty honeybee hives? Similarly, imagine a small Hebridean island called Oronsay, supporting a population of the endangered moss carder bee. Would this be a sensible place to put another twenty honeybee hives? In both cases (one theoretical, one real), the obvious answer would be no. But it is surprising how much hot water I have got into for saying so.

I have digressed. Jane, Andrea and I set out to discover whether bumblebees in Tasmania were doing any harm. So far as native bees were concerned, the answer seemed to be no. We also looked at the seeds of the introduced tree lupins to see whether the arrival of bumblebees had awakened a new sleeper weed. In this respect, bumblebees appeared to be having a strong effect. Lupins are related to peas, and their seeds form in rather similar pods. Each yellow spike has lots of individual flowers attached to a central stem, and if all goes well then each flower produces a pod full of seeds. In places where there were no bumblebees, roughly 60 per cent of flowers fell off without setting any seed, and where they did set seed there were only about two seeds per pod. In contrast, in sites where bumblebees had become common, only 30 per cent of flowers set no seed and on average there were about six seeds per pod. Overall, bumblebees were allowing each lupin plant to produce more than four times as many seeds. It seemed likely that the arrival of bumblebees might well lead to tree lupins becoming as big a weed in Tasmania as they are in New Zealand.

Eleven years later I wanted to find out what had happened. By that time I was based at Stirling University, and an excuse to escape the Scottish winter for some Antipodean sun was even harder to resist than ever. So it was that in December 2010 I found myself reprising our whistle-stop tour of Tasmania, this time with a PhD student named Ellie Rotheray, who was moonlighting from her studies of the UK's rarest fly, the pine hoverfly.* On our first morning, with lashings of coffee inside us to counteract the effects

* This wonderful beast lives for most of its life as a 'rat-tailed' maggot in a puddle of rainwater formed in the heart of a rotting pine tree stump, and nowhere else. The maggots have a telescopic breathing tube attached to the rear end which earns them their unappealing name – as maggots go they are actually rather cute. Sadly there seems to be only one tiny population of this fly left in Britain, but Ellie has been busy breeding them in captivity and chopping holes in pine stumps elsewhere to release them into.

of jet lag and a thirty-eight-hour journey, we tried to find some of our former lupin populations just south of Hobart. We hadn't had the benefit of a GPS back in 1999 so we had no precise coordinates to go to, but I did have scribbled notes on a very tatty map and I thought that I remembered the sites fairly well. The first lupin patch on the edge of the pretty seaside town of Kingston was still there, just, but had largely been destroyed by diggers creating what appeared to be a rough car park. Only a couple of bedraggled plants survived, half-buried under the rubble. The second patch, a little further south along the side of a main road, seemed to have disappeared entirely. We drove up and down looking for it, as I scratched my head and mumbled apologetically, 'I'm sure it was round here somewhere . . .' I was starting to fear that we had flown 11,000 miles on a wild goose chase. We headed south, and as we failed to find a couple more lupin populations, I felt increasingly foolish. It wasn't until we reaching the sleepy fishing village of Dover, near Tasmania's southern tip, that we saw our first decent patch of yellow lupins. Dover is a tiny place, a cluster of fibreboard houses strung around a beautiful sandy bay, the sand strewn with abalone shells from the local fishery. At the back of the beaches the lupins had been spreading in the dunes, forming a dense strip for half a mile along the bay. They looked stunning, framing the icy blue sea, surf and sand with a crescent of vivid yellow, but they had been steadily strangling the native flora, forming a dense, impenetrable thicket.

The pattern was similar elsewhere: coastal populations tended to have expanded, while the inland populations tended to be small and ephemeral. Many of the inland populations we had found in 1999 had gone, wiped out by herbicides or development, but other populations had popped up here and there. The worst infestation was on Bruny Island, a gorgeously remote and sparsely populated twist of land off the east coast of southern Tasmania, home to more echidnas and fairy penguins than people. Here, the lupins were running amok, spreading along the sandy coast and in dense swathes into the gum forests inland. We dutifully counted every plant in all the lupin patches we could find around Tasmania; as ways to earn a living go, I can think of many worse for it is one of my favourite places on earth, paradise for a naturalist. In our twelve-day tour, as well as lots of lupins we saw pods of dolphins frolicking just feet from the shore, sea eagles, flocks of black cockatoos, penguins, duck-billed platypus, wombats, more echidnas than you could shake a stick at, and even a beer-drinking pig named Priscilla.*

The bumblebees seemed to have taken very well to Tasmania, having now spread throughout the island from the wild and windswept west coast to the sunny north-east. This despite recent genetic evidence suggesting that they were all the descendants of a very small number of queens that made it here in 1992 – perhaps just one or two – and hence all horribly inbred.

What we didn't see this time was Tasmanian devils. Whilst the roads were still as littered with corpses as they were in 1999, where previously the devils had comprised perhaps a quarter of the body count, now there were none. The poor brutes have been afflicted by a grotesque plague, a deadly facial cancer which is infectious and spreads when they bite each other – which being rather bad-tempered beasts they are prone to do. No one knows where it came from, but it is most likely a disease accidentally introduced by man. It has all but wiped them out, so perhaps I will never now see one in the wild.

* A must-see in northern Tasmania, Priscilla is to be found at the Pub in the Paddock in the remote village of Pyengana. For $1 you can buy a watered-down beer which she will enthusiastically guzzle from the bottle while grunting contentedly.

Once back in a frightfully cold Scotland (the temperature in December 2010 regularly fell to -20°C), I set about analysing the data. Overall, we found 76 per cent more plants in 2010 than in 1999, with many coastal populations having doubled or tripled in size. The conclusion is that it does look as if lupins in Tasmania are spreading and likely to become a major problem in the future.

In recent years the Australian horticultural industry has also been making applications to allow the release of bumblebees on the mainland. No doubt the Australian tomato growers are desperate to put down their vibrating wands and let bumblebees do the work, and it is hard to blame them for this. It would certainly save them money, and maybe they would get bigger, sweeter tomatoes too. But my guess is that the cost to other farmers through worsening weed problems could vastly outweigh the benefits to the tomato industry. Thankfully and wisely, the Australian government have turned down the applications so far. The country doesn't need any more non-native species. Nonetheless I worry that bumblebees may one day soon mysteriously appear there, much as they did in Tasmania. After all, the distance from Tasmania to Victoria, the neighbouring state on the mainland, is much smaller than that from New Zealand to Tasmania, and regular passenger ferries cross between the two.

I love bumblebees. Beekeepers love their honeybees. Both are enormously valuable and important creatures. But mankind has wrought enormous harm on our ecosystems by shifting species around the globe. In New Zealand, I enjoyed watching rare UK bumblebees and honeybees happily feeding upon and pollinating huge, colourful stands of viper's bugloss, lupins, foxgloves and clover. Yet as I stood there contentedly chewing upon a pie made of venison (another non-native), I knew that I was watching an ecological travesty. The truth is that in New Zealand we have patched together a Frankenstein ecosystem on the wrong side of the world, and that in so doing we have annihilated the native creatures that used to live here. We all have to accept that, in the wrong place, both bumblebees and honeybees can do harm, and that very great care should go into considering the risks before any more bees are released outside their native ranges.

CHAPTER EIGHT

Quinn and Toby the Bumblebee Sniffer Dogs

The music of the busy bee
Is drowsy, and it comforts me;
But, ah! 'tis quite another thing,
When that same bee concludes to sting!

Andrew Downing
(nineteenth-century American horticulturalist)

One of the great difficulties in studying bumblebees is finding their nests. They can be in all sorts of odd places, many of them tucked underground in old rodent burrows, in hedge bottoms or amongst the roots of a tree. Others prefer compost heaps, bramble thickets, lofts, rockeries, holes in trees or tit boxes. All that these sites have in common is that they tend to be tucked away out of view.

A honeybee or wasp nest can contain tens of thousands of workers, and the traffic passing to and fro becomes pretty obvious if the nest is in a garden or other place frequented by people. All you have to do is look for streams of flying insects and follow them back to their nest. The number of bees in a bumblebee nest increases through the spring, and the size to which such nests grow varies between species, but even the largest ones rarely reach as many as 300 or 400 workers. For most of the year, the traffic amounts to no more than one or two bees a minute, so even garden nests are easily overlooked. In my experience most gardens have several bumblebee nests each year – at the time of writing my quarter of an acre in Dunblane has at least two: a small buff-tailed nest in an old compost heap, and a very large white-tailed nest under a piece of old wood beneath the children's trampoline. Indeed my eldest two boys, Finn and Jedd, had been happily bouncing about just above the nest for several weeks before I pointed it out to them. The bees themselves seemed not the slightest bit perturbed by their trampolining.

All this poses a problem, for finding a nest is enormously time-consuming, and finding the nests of rare species is even worse. Because so few rare ones have ever been found, we know almost nothing about where they might occur, providing a bit of a catch-22. With rare species such as the short-haired bumblebee or the ruderal bumblebee, most of our knowledge comes from the work of Frederick Sladen 100 years ago (they weren't so rare in his day). A detailed understanding of their nesting ecology would be really useful if we wanted to provide them with extra nest sites to help boost their populations. If we could find wild nests, we could study them, and provide a window on aspects of their lives that remain obscure. For example, we could catch the bees coming back to the nest and see what pollen they were carrying, and so learn about where they were collecting their food; or we could study what creatures attack the nests, or count the nests to see how many there are in different parts of the country or in different habitats. All of these things would be valuable to finding out more about the ecology of bumblebees, understanding why they are declining, and working out how best we can help them. So all in all, it would be jolly handy if we could come up with a reliable way of easily finding lots of bumblebee nests.

Bumblebee folk have been pondering this issue for years. One solution is to recruit thousands of people to help. In 2004, Juliet

Osborne, the scientist who carried out the bee radar experiments, advertised for members of the public to take part in a national bumblebee nest survey. Her volunteers were then asked to take a deckchair and sit in their garden for twenty minutes. (The deckchair wasn't compulsory but it made the whole exercise much more relaxing – as did a cold gin and tonic.) During this time they were asked to stare at a selected 6-by-6-metre area of their garden, watching for bumblebee nest traffic. Although I have said that such traffic is rather slow and hard to spot, if you stare at a patch of ground for long enough, then if there is a nest entrance you will eventually notice bees coming in and out – and any bumblebee flying out of a hole, or down to ground where there are no flowers, is likely to indicate a nest. The rationale was that twenty minutes would be long enough to spot a nest, though it may seem like an awful long time when nothing is happening (hence the importance of the chair and the gin). If a nest was spotted, the volunteer was then asked to identify the species as far as possible, although most volunteers were only able to classify the bees as belonging to one of five colour groups (e.g. all-brown bees, or bees with two yellow bands and a white bottom, etc.).

Having studied their gardens, these volunteers were asked to repeat the exercise in one countryside habitat, chosen at random from a range of options. Once they had done both, they completed forms describing the places they had watched and any bumblebee nest they had found. Impressively, 719 volunteers from all over the country stepped forward, completed the exercise, and returned their forms to Juliet. Between them, they had managed to find 215 bumblebee nests. Of course this means that over two-thirds of the volunteers had spent forty minutes staring at the ground and found nothing. Disappointment aside, however, thankfully they still returned their data, for in science, the zeros are just as important as the more exciting positive numbers.

One of the more striking results to emerge was that our gardens seem to house more bumblebee nests than the countryside, for Juliet found that, on average, there were thirty-six bumblebee nests per hectare of gardens, with much lower densities in farmland. I find this both encouraging and depressing – encouraging in so far as gardeners such as me who try to make their gardens as wildlife-friendly as possible are clearly doing something right. Our gardens provide lots of good places for bumblebees to nest – old compost heaps, sheds and patios to nest under, rockeries full of cavities and so on – along with a huge variety of flowers; and although individual gardens are very variable in how many good bee-flowers they have, bees are of course no respecters of boundaries and workers from a single nest will range over hundreds of gardens in search of food. At any one time in the spring and summer at least some of those gardens are bound to have some rewarding flowers. By contrast – and here is the depressing bit – farmland has rather fewer places for bumblebees to nest, and very often startlingly few flowers. Arable fields cover much of lowland Britain and they provide no nest sites for bumblebees – being regularly tilled, they have no old rodent burrows. With modern farming techniques, arable fields also tend to be largely free of weeds, so there are no flowers (unless they are sown with a flowering crop such as oilseed rape, which does provide lots of flowers for a couple of weeks but then none for the rest of the year). Juliet's survey did find that hedgerows and fence lines are the best places to find bumblebee nests in farmland, but there are far fewer of these than there used to be because fields have got much bigger.

Following on from Juliet's work, one of my PhD students, Gillian Lye, has also enlisted the public's help in sending her records of the bumblebee nests in their gardens. In this informal study, people are simply asked to fill in a questionnaire if they happen to find a bumblebee nest. This more haphazard approach has so far recorded 519 nests, once again showing that gardens provide a diversity of good sites. Tit boxes appear to be particularly popular, especially for the early and tree bumblebee species. Moreover, some bumblebees seem to like nesting beneath or within certain materials: red-tailed bumblebees, for example, tend to nest under stones or patio slabs.

Gillian and Juliet's surveys have produced some really interesting results, but they are very biased towards what is happening in gardens. They almost never turn up information on nests of rare bumblebees, probably because our rarest species tend to live in such unpopulated corners as the Hebrides or on Salisbury Plain.

An alternative approach to finding bumblebee nests might be to lure them to nest in artificial boxes. Most garden centres sell such boxes and, if these were readily used by bumblebees, they would provide a straightforward way to obtain nests for study. During her PhD, Gillian tested a range of commercial bumblebee nest boxes, and also a selection of home-made designs. Gillian tried out well in excess of 500 nest boxes during her three years, scattering them in different habitats, above and below ground, with or without entrance tunnels, and so on, and then monitoring them regularly to see which ones were occupied. Her prodigious efforts were poorly rewarded: about half a dozen boxes in total were occupied. That is to say, half a dozen were occupied by bumblebees. Almost all housed spiders, slugs, earwigs, woodlice, wasps and so on. If they were rebranded as woodlouse boxes they could be regarded as highly successful, but as bumblebee nest boxes they represent a poor investment.

Over the years, I have hatched various schemes to find bumblebee nests, some more ridiculous than others. I have tried catching queens that were collecting pollen (they do this only once they have established a nest) and tying pieces of silver tinsel to them. My hope was that this would both slow them down and make them more obvious, so that I could run after them back to their nest. It didn't work; the bees either flew far too high and fast to follow or, when I experimented with larger pieces of tinsel, simply flopped down to the ground and devoted themselves to biting it off. I have also tried plotting the direction in which worker bees leave flower patches, in the hope that I could discern a pattern which I could then follow to find their nests, but bees leaving flower patches seem to show no pattern in the directions in which they head off. Another suggestion involves the infrared imaging that the army uses to spot the enemy at night, for as bumblebee nests are warm an above-ground nest should show up even when hidden under leaves. Sadly the equipment involved costs hundreds of thousands of pounds so we have yet to try.

It has long been known that one animal has no trouble in finding bumblebee nests: the badger. Badgers are voracious predators, digging up and consuming everything that they find – bees, grubs, wax, honey, the lot. They must get stung, but they seem not to care, or perhaps the honey makes the pain worthwhile. I once carried out an experiment to see how well bumblebee nests fared on different farm types compared to gardens, in which I placed artificially reared bumblebee nests in various locations. This worked pretty well; the nests in gardens tended to grow much more quickly and become heavier than those on farms, presumably because they have more flowers. The only problem I had was that quite a few of my nests (both in gardens and on farms) were eaten by badgers, leaving nothing but chewed, tattered and very empty boxes.

Salisbury Plain is a magical place, a glimpse of what much of the English countryside may once have looked like – vast tracts of flower-rich downland, supporting all sorts of rare wildlife including perhaps the richest bumblebee fauna in the UK. The Plain escaped agricultural intensification only because the army began buying up the land in 1897 and now owns 38,000 acres – the largest remaining patch of flower-rich chalk grassland in north-west

Europe. Of course the peace is occasionally shattered by a barrage of artillery shells or the deafening clatter of a passing tank, but for the most part the wildlife here is free to thrive. The Plain harbours a substantial badger population, and on many fieldwork expeditions I have found signs of recent excavation, holes dug in the ground with a few forlorn bees still hanging around and telltale paw prints in the loose soil. It is said that badgers turn to eating bumblebee nests in dry summers when the worms that they eat for much of the year burrow too deep for them to find. Of course badgers are nocturnal; they have poor eyesight, and clearly find the bumblebee nests by smell. In fact the nests are especially smelly, at least when reared in captivity; I have heard the smell described as like Christmas cake, but if so it is not a type of Christmas cake that I would like to eat. You can create a similar odour by pouring black treacle and sherry over a pair of dirty running socks, sealing them into a Tupperware box and then leaving it in a warm place for a month. Not that I have tried this, of course.

So the solution may well be a pet sniffer badger. Sadly for me, however, badgers do not domesticate well, remaining grumpy and rather dangerous. But if not a sniffer badger, how about a sniffer dog? Dogs are well known for their incredibly sensitive noses; after all, once trained, they can sniff out drugs, explosives and even banknotes. In the USA, dogs have been trained to find termite infestations in houses long before any external signs of damage are visible.* Dogs have even been used successfully to detect cancers

* Termites are wood-boring insects that can destroy a timber-framed house in no time at all. They are fascinating creatures, not least because they are rather like miniature cows – just as a cow has a rumen, the stomach in which it digests the cellulose in grass by fermenting it in a warm broth of bacteria, so termites have a 'paunch', a special stomach in which they digest the cellulose in the timber they consume.

in humans. If they can manage all that, then surely they could sniff out a highly odoriferous bumblebee nest? I chatted about this idea with Juliet and some of my research group over a number of years, but never quite found time to follow it up. Then, one day in 2004, one of my more go-getting PhD students, Ben Darvill (who subsequently went on to help me found the Bumblebee Conservation Trust), came up with a telephone number for the Defence Animals Centre. Based in Melton Mowbray in Leicestershire, this organisation trains all the army dogs used to detect explosives in Iraq and Afghanistan. They also produce dogs for customs and police use. If anyone could train up a bumblebee sniffer dog, then surely it was them?

I must admit that I felt like a bit of a fool when I rang, fully expected them to laugh, or put the phone down thinking my call was a hoax. Compared to the serious business of sniffing out bombs, asking them to train a dog to sniff out bumblebee nests seemed rather frivolous. To my surprise – and their immense credit – the staff at the Defence Animals Centre were immediately interested. I think perhaps they were a little bit bored of training dogs to do the same thing time after time, and they fancied a challenge. Whatever their reasons, they invited us to visit. So it was that Ben and I and another PhD student named Joe Waters drove up from Southampton on a rainy day in November. I was most disappointed that I didn't see a single pork-pie shop on the way through Melton Mowbray. The Defence Animals Centre's huge site, which also houses the Royal Army Veterinary Corps, has vast stables, along with rows of dog kennels from which a constant howling could be heard, and occasional office blocks scattered across a slightly bleak hillside. As we drove in we passed immaculately groomed and impeccably behaved cavalry horses that were being put through their paces in the damp autumn air, their breath creating clouds of steam.

We met Dutch, a senior dog trainer, who took us for a demonstration. We drove behind his Land Rover along various muddy tracks until we came to a large semi-derelict army building, which might have once been a mess hall and barracks, but clearly hadn't been used for years. There Dutch opened the back of his Land Rover and out bounded a very excited black Labrador called Miffy, eager to get to work. One of us held the dog's lead while Dutch entered the building and hid an object. Then he led us all into the building and gave Miffy the command to start work. She dashed around, her tail wagging crazily, jumping on furniture, crawling under cupboards, poking into every nook and cranny and snorting when she got dust up her nose. After about two minutes her seemingly random searching brought her to a small cupboard set against a wall, at which point she dropped to the ground, her whole body quivering with excitement, nose to the cupboard, her tail a rigid point behind her. Dutch opened the cupboard and there inside was a tiny pea-sized piece of cocaine wrapped in cling film. He threw a tennis ball to Miffy (her reward for success) and she bounded off with it, looking very pleased with herself. All most impressive. If Miffy could be trained to find such a small sample of drugs wrapped in cling film, then surely a dog could be trained to find a bumblebee nest? We had brought with us an old, frozen bumblebee nest in an ice box for Dutch to use as training material. It really was quite smelly and he agreed that it ought to be easy enough to train a dog to find it, but he explained that it might take a few months for them to locate a suitable animal. So we left him with the nest and returned to Southampton.

As it turned out, his call was a long time coming. It seems that finding good sniffer dogs is not simple, and perhaps finding one for bumblebee nests was not their top priority, what with wars in Afghanistan, Iraq and so on. The Defence Animals Centre uses a wide range of breeds, obtaining their dogs from rescue centres, but the first few that they tried to train to find bumblebee nests were deemed unsuitable after just a few days. In the spring of 2005 they settled on a golden Labrador with the slightly unfortunate name of Chad. He showed early promise but then after a few weeks he too was rejected; apparently he lacked focus and was too easily distracted. Dutch was also concerned that, unless a sniffer dog was especially obedient, it would soon shove its nose into a real bee nest and get stung, which might very well put it off from ever finding a bumblebee nest again. So he wanted a dog that would concentrate and always keep a little distance from the source of the smell. A suitable replacement for Chad was not forthcoming, and before we knew it the 2005 bumblebee season was coming to an end and we still had no sniffer dog. Efforts were abandoned until the following spring. Finally, in March 2006, the DAC recruited Quinn, an English springer spaniel. He, apparently, was the real deal, and after two months of training he was ready for action. When Joe then agreed to become his handler, he went back to Melton Mowbray to be taught how to look after Quinn and keep him trained.

Joe's PhD was on the ecology of Hebridean bumblebees. We hoped that with Quinn's help, he would be able to find dozens of rare bumblebee nests. The Hebrides are one of the last strongholds of the great yellow bumblebee, Britain's rarest species, and only a handful of nests of this species had ever been found. Before heading off to the Hebrides, Joe put Quinn through his paces; in exchange for a pint, he persuaded a friend to bury small samples of nest material from a range of bumblebee species. It was important that Joe didn't do this himself as he might have then subconsciously helped Quinn to find the samples. The bits of bee nest were placed in perforated plastic pots so that the smell could escape. Empty 'control' pots were also buried. In these trials Quinn worked fantastically – he found every pot containing bee nest, showing no interest at all in the control pots. It looked as if we might have finally cracked it.

So it was that Joe and Quinn spent the summer of 2006 on the lovely island of Tiree in the Inner Hebrides, living in a camper van and searching for rare bumblebee nests. Tiree is a lovely little island, with superb shell-sand beaches, lots of flowers, and heaps of interesting bees. Although it was hard to argue with his choice of field site on academic grounds, I always suspected that Joe chose Tiree in part because it has some great surfing breaks. In their summer of searching, Joe and Quinn found twenty-five nests of the moss carder bumblebee and four of the great yellow. This was pretty good, and revealed some interesting things. Joe and Quinn discovered that these bees often nest very close together; they don't seem to mind if the entrance to another nest is just a few feet away from their own. Great yellow bumblebees, it seems, love to nest in old rabbit burrows, and the majority of bumblebee nests that Joe and Quinn found were either in sand dunes or on the sandy 'machair' inland from the dunes.* Yet although this sounds impressive, the truth is that Joe and Quinn were only averaging one bumblebee nest every couple of days, whereas I'd been naively hoping that they might find dozens a day. This said, I had little idea how many nests there were on Tiree for Quinn to sniff out, so I didn't know whether to be pleased or disappointed.

By this stage the bumblebee season was also drawing to a close, and Joe had to keep Quinn trained over the winter when there

* Machair is a rare and very beautiful habitat found only in the west of Scotland and Ireland. It is made of flat plains of wind-blown shell-sand upon which grow the most stunning swards of flowers. It is now a last refuge for a range of rare creatures such as corncrake and the great yellow bumblebee, which have been unable to cope with farming changes on the mainland of Britain.

are no wild bumblebee nests to find. Moreover, a sniffer dog needs constant practice, so every day Joe had to thaw out bits of frozen bumblebee nest, bury them, and get Quinn to find them. In the early spring of 2007 Quinn and Joe then returned to the Defence Animals Centre for an update on their training. Declared fit for operation, they were unleashed once more on to unsuspecting bumblebee nests, this time in Hertfordshire, as part of a big project designed to find out how fields of oilseed rape and beans affect bumblebee nests. Most arable crops are of no interest to bees (think cereal fields, vast flowerless wastelands from a bee's perspective), but rape and beans are both popular with bees and benefit from bumblebee pollination. You might therefore assume that having huge fields of flowers must be good for bees, but these crops flower for only a few weeks and the debate is whether such a brief glut of food is a good or a bad thing. So the plan was for Joe and Quinn to find us bumblebee nests near fields of flowering crops, and a similar number away from them, and then to monitor how they all fared through the season.

Quinn found Hertfordshire hard going. We already knew from Juliet's nest survey that most bumblebee nests in arable farmland were likely to be in thick hedges and woodland edges, and compared to the open machair and sand dunes of Tiree, Quinn couldn't easily get into these to sniff about. We began to suspect that many of the nests we sought might be in bramble thickets and at the bottoms of the hedgerows where the vegetation was too impenetrable for Quinn. Whatever the reason, he found far fewer nests than we had hoped. Even worse, the nests he did find were all soon dug up by badgers. The plan was to go back to each nest every week and count the bee traffic coming and going to give us some idea of how big each nest was. Sadly, the dozen or so that Quinn managed to find were all dug up by badgers within the following week. It was almost as if the badgers were following

Quinn's scent, or perhaps both Quinn and the badgers were finding the smelliest nests. Maybe the disturbance of Quinn trampling around near the nests made them easier to find, but whatever the reason, this was pretty disastrous for the project. We had no nests left to follow and see what effects flowering crops had on them.

Disillusioned by the lack of success, Joe now decided that bumblebee research was not for him and announced that he was going to retrain as a teacher. I couldn't really blame him – when a project isn't going as planned it can be enormously frustrating. In any case an academic career is uncertain and poorly paid, and there aren't anywhere near enough jobs for most PhD students to be able to stay in research once they finish. The trouble was that Joe and Quinn were an inseparable team, and although we had paid the DAC to train Quinn I didn't think that I could ask Joe to hand him over when he left. And so it was that we found ourselves once more without a sniffer dog. Quinn's early successes on Tiree were sufficiently promising that it seemed a terrible shame to give up on this idea. On the other hand, we didn't have anyone willing and able to take on the full-time task of becoming the handler of a new dog, even if we had one.

It was around this time that I first met Steph O'Connor. Steph had applied to study for a PhD with me, and although she didn't make my initial shortlist she subsequently rang me up and begged for an interview. Impressed by her enthusiasm, I thought I'd give her a chance. Ben Darvill was on the interview panel, and his favourite question is to ask the candidates whether they have ever found themselves in a particularly challenging or stressful situation, and if so, how they coped. The candidate who eventually got the PhD, Nicky Redpath, had described how she and a friend had been held at gunpoint whilst on holiday in Kashmir, and had somehow talked their way out of it. Exciting stuff, but nowhere near as bizarre and amusing as Steph, who described how she had come across an advert on the Internet from a Swedish man who wanted a girl willing to dress up in Viking costume and help him stage historical re-enactments. When Steph volunteered and went to stay in his house, however, she soon found that his main interest was in getting her out of her traditional Viking dress. Instead of returning to Britain, and demonstrating a typically cavalier attitude to personal safety and common sense, Steph stole his dog – which he wasn't looking after properly – and went to live somewhere in the woods, in a tent, with the dog and two young English guys who professed to be Nazis. Steph recounted this tale (which went on further, but I will spare her blushes) in her rather posh voice and with a beaming smile, and I and the rest of the panel were aching with laughter by the end of it. Of course this didn't necessarily make Steph the ideal person for a PhD, but it certainly made her interview memorable. And so it was that when, a few weeks later, I found myself with some money to employ a temporary research technician, I couldn't resist offering the job to Steph.

Soon after she started work it became apparent that Steph would be a perfect handler for a new sniffer dog. If ever there is some kind of global catastrophe, Steph will undoubtedly be among the survivors. Her idea of a perfect weekend is to go rabbiting with her three pet ferrets, or out in the woods shooting pigeons, or making home-made wine out of unpromising root vegetables. On one occasion she came in to work with a packed lunch of squirrel casserole; at the time her skinning skills were not great and she had to spit out clumps of fur as she ate. She assures me that the casseroles have improved.

When I asked the Defence Animals Centre if they could train up a second bumblebee sniffer dog, to be handled by Steph, the Leverhulme Trust kindly agreed to provide funding for three years, thereby assuring her salary. This time the DAC found a suitable dog swiftly, a springer spaniel named Toby. Once he was ready,

Steph went down to Melton Mowbray to learn how to handle him, and by the spring of 2008 the two were ready for action. Steph was keener on publicity than Joe had been, so we put out a press release and she and Toby ended up appearing on BBC breakfast television and on *The One Show*. Toby quickly became something of a minor celebrity, and had many invitations to visit schools. His sniffing abilities are quite phenomenal, and he gives a great demonstration for kids. If he is making an appearance in a school, we post them a few strands of moss from a bumblebee nest in advance and ask them to hide them somewhere in the school grounds. To the kids' delight, Toby dashes around excitedly, invariably finding the nest material within minutes.

Frustratingly, however, and just like Quinn, Toby is nowhere near as good at finding real, live, wild bumblebee nests as he is at locating the bits of nest that we hide. Although he finds some, he scampers right past others. We suspected at first that freezing might slightly change their scent, and to overcome this, Steph retrained him using fresh nest material. When this didn't seem to help, we wondered whether nests of different bee species might smell different, so that Toby would learn only to find nests of whatever species he had recently been trained to find. In fact, however, he finds nests of a range of species. Our latest theory is that perhaps real nests containing live bees smell different from bits of empty nest, but it is difficult to use nests with real bees in them when training Toby as the bees quickly fly away. At the time of writing we are trying to analyse the chemical constituents in the hope that this might help us to understand why Toby can smell out some nests but seems unable to detect (or perhaps chooses to ignore) others.

Steph has also compared Toby's nest-finding abilities to those of human volunteers. Juliet Osborne's approach of asking volunteers to stare at a fixed area of ground for twenty minutes does seem to work, but the majority of the time there is no nest in the area being surveyed, so it is pretty boring. An alternative tactic is to allow human volunteers to roam free when searching, which is more fun. To compare them to Toby, Steph asked dozens of volunteers from Stirling University to search the woodland on campus for bumblebee nests. Each person did one twenty-minute 'fixed search' (staring at a randomly allocated patch of ground), and one twenty-minute search where they could walk wherever they liked. Steph shadowed them as they went, and she also searched the woodland with Toby. Tragic as it is to report, it turned out that, despite his army training and sensitive nose, Toby is no better than a novice human volunteer in terms of the average number of nests he can find in twenty minutes. In a fixed search, volunteers had roughly a one in nine chance of seeing a bumblebee nest. When allowed to wander about, they had on average a one in four chance of locating one. Toby's rate was also one nest per four twenty-minute search periods. Experienced bee researchers fared no better than novices (I was one of the volunteers, and to my frustration I found nothing).

This is not particularly good news for Toby and the future of sniffer dogs in bumblebee research. Given that a dog requires months of costly training, then a full-time handler, and needs constant practice with bits of buried nest over the winter, he needs to be substantially better than a human to justify the daily tin of Pedigree Chum. If we cannot find a way to improve his skills, then Toby may soon be putting on his slippers and lighting up a pipe.

Ironically, Steph herself has become very good at spotting bumblebee nests. When shadowing the volunteers she spotted many nests that they did not, and often when working with Toby she finds the nests before he does. After two and a half years of searching, she has become something close to the finely honed bumblebee nest-finding machine that we had hoped Toby would be. Between the two of them they have managed to track down over 100 bumblebee nests in the last two years, and Steph has been studying them in detail to find out what their main natural enemies are. It may be that there is no magical answer to locating nests beyond persistence, a quality that she has displayed in spades.

CHAPTER NINE
Bee Wars

Is this wretched demi-bee,
Half-asleep upon my knee,
Some freak from a menagerie?
No! It's Eric the half a bee!

Monty Python, 1972

Bumblebees are surely among the most gentle and friendly of insects. When visiting flowers in the garden they are placid and simply fly away if disturbed by a human or another bee. Unlike wasps or honeybees, most bumblebees don't even seem to mind very much if you poke around their nest, stinging only as an absolute last resort. Moreover, they are highly social creatures, with the daughters working together with their mother to look after their young and to gather food. Philosophers and writers from Aristotle and Plato to Shakespeare and Marx have used bee societies as a model example against which humans are regarded as comparing poorly. After all, what could be more harmonious than a sisterhood of celibate bees devoting themselves to helping their aged mother and their younger sisters? Yet this apparently altruistic and idyllic nunnery is not what it seems, for within the shadowy confines of the nest violent fights do occur, and cannibalism, infanticide and murder are rife.

To explain this dark side, it is first essential to look into why bees are normally so sociable, and this is a little complicated. In most creatures, parents look after their offspring because their offspring carry their genes into the next generation. Parents that leave lots of offspring behind pass on many genes, so any gene which makes a parent good at producing and rearing offspring will become more and more common in successive generations. Self-evidently, the genes of parents that leave few offspring will quickly disappear. This is the basis of evolution by natural selection (of course you do not need to know or understand this to be a good parent). Some evolutionary biologists even argue that we are simply vehicles manufactured by our genes as mechanisms to help them multiply – a somewhat disconcerting thought.

At the risk of this sounding a little like a textbook, I need to explain a bit about genes and inheritance. These two words strike fear into the hearts of biology undergraduates, for they associate them with fiendishly complex exam questions such as, 'Calculate the probability that the child of a left-handed colour-blind woman from Cardiff married to a one-legged Glaswegian with sickle-cell anaemia will have brown eyes and a limp.' I will keep this as simple as possible and there is no exam at the end. Genes are carried in chains (called chromosomes), contained within the nucleus of almost every cell in our body. Most animals, including ourselves and female bees, are diploid, meaning that we have two copies of each chromosome, and hence two copies of each gene. In humans, we happen to have twenty-three pairs of chromosomes. Female bumblebees have between twelve and nineteen pairs, depending on the species. Oddly enough, adder's-tongue fern holds the record at over 1,200 pairs, although why this rather nondescript little plant needs so many is unknown. These chromosome chains contain all the information needed to build a fully functioning human, bee or fern, a bit like a vast instruction manual.

Each gene can be seen as a recipe – they provide the information required to build a particular protein needed in the body. It is handy that we have two copies of each gene, for some are duds – the recipe contains a mistake, and so does not work. For example, roughly one in twenty-five Caucasians has a mistake in the gene which provides the recipe for a protein with the snappy name of cystic fibrosis transmembrane conductance regulator, or CFTCR. So long as we also have a good copy of this gene, we are fine. If by chance we have two duff copies, we have cystic fibrosis.

When we produce offspring, we pass to them one of each pair of chromosomes, and hence one copy of each gene, good or bad; they obtain the other copy from their other parent. To do this, we have a special type of cell division (known as meiosis), which takes place in our gonads whereby normal, diploid cells divide to produce gametes – sperm or eggs – which are haploid, meaning they have just one copy of each of the twenty-three chromosomes. During sexual reproduction, gametes from each parent fuse to produce a diploid cell, a zygote, which then divides and grows to produce a new organism. It follows from this process that each of your offspring carries 50 per cent of your genes (the remaining 50 per cent coming from their other parent). In evolutionary terms, your genes have broken even if you have two children, for on average each of your genes will have been passed on once. More than two children, and your genes might consider that they have done well for themselves. Fewer than two, and your genes might rightfully be disappointed in your performance.

In humans, one of the pairs of chromosomes determines sex; the sex chromosome comes in two types, X and Y. Your mother had two X chromosomes. Your father had an X and a Y, and your sex depends solely on whether he passed on to you his X chromosome or his Y chromosome. Under this common genetic system, you are not only 50 per cent related to your offspring, but you are also 50 per cent related to your parents and to your siblings. By extrapolation, you are 25 per cent related to your grandchildren, grandparents, aunts, uncles, nephews and nieces, and so on. The same applies to most animals, but not to bees. In bees, it is much more complicated.

Bees belong to the Hymenoptera, a huge and very successful insect group that also contains ants and wasps. It is not by coincidence that the Hymenoptera includes most of the known social insects. It is because of their rather weird genetics. In bees, sex is determined by a single gene. If an individual has two different copies of this gene, it is female. If it has two identical copies, or just one copy, it is male. Female bees, like us, have two copies of each chromosome. Male bees, typically, have just one. To produce a son, a female bee has just to lay an unfertilised egg; the haploid gamete develops into a healthy son. Sons have no father (male bees are bastards). To produce a daughter, she fertilises her egg using sperm from a male; in bumblebees this sperm had been stored inside the queen since the previous summer. So long as the copy of the sex-determining gene in the sperm is different from each of the two different copies held by the mother, then these diploid offspring will all be female. In a normal, healthy bee population there are dozens (perhaps hundreds) of different versions of this gene, so it is unlikely that the gene of the father will match either of the versions held by the mother.

One odd consequence of males being haploid is that they do not need the process of meiosis to produce haploid gametes; all of their cells are haploid, and hence all of their sperm carry exactly the same set of genes.

My apologies if you are on the verge of falling asleep. I teach this stuff every year to the third-year students at Stirling, and every year I notice that at least half the class have tuned out within five minutes, no matter how much I jump about and try to make it sound exciting. The importance of all this is that it results in some very odd patterns of relatedness. Daughters get one copy of each chromosome from their mother and one from their father, so they are 50 per cent related to their mother, and she to them. Sons get one copy of each of their mother's chromosomes, so they carry 50 per cent of her genes. Sisters must share identical genes from their father's side, and on average share 50 per cent of the genes they get from their mother; hence overall they are 75 per cent related. This is a crucial point – sister bees are more closely related to each other than they are to either their mother or their own offspring. Another strange quirk of this system is that a father is 100 per cent related to his daughters (she has all of his genes), but she is only 50 per cent related to him (half of her genes came from her mother).

If by now you are utterly confused, as I was when I first tried to get to grips with this, don't worry. All that you really need to remember is this: a female bumblebee is 50 per cent related to her daughters and sons, but 75 per cent related to her sisters. Now, why does any of this matter?

I started this slightly tedious discourse on bumblebee genetics by saying that parents look after their offspring because their offspring carry their genes. Now, put yourself in the (six very small) shoes of a worker bee in a bumblebee nest. She could try to lay her own eggs. Because she has not mated, these will be sons, which carry 50 per cent of her genes. Even supposing she had been able to mate and could produce daughters, they will also carry only 50 per cent of her genes. Alternatively, she could help to rear sisters, which carry 75 per cent of her genes. So all else being equal, the best way to increase the number of copies of her genes is to rear sisters rather than her own offspring. This, in essence, is why highly social behaviour has become common in the Hymenoptera, but is rather rare in most other organisms; their odd genetics have predisposed daughters to help their mother rear their sisters rather than trying to reproduce themselves. An ant, wasp or bee nest is a vast, tightly knit group of closely related sisters helping their mother to produce more and more sisters.

This, hopefully, explains why a bumblebee nest is so harmonious; so long as they are rearing their sisters, the worker bees should be content. They have no incentive to try to have their own offspring. The problem comes when it is time to produce sons. Somebody has to produce them; in bumblebees, the nest will die in the winter, so the only option is to make new queens and for the males to mate with them before summer's end, so that the mated queens can survive the winter. Hence, in high summer, the queen bee starts laying both fertilised (female) and unfertilised (male) eggs. Up to this point she has been releasing a pheromone signal, instructing her female offspring to develop as workers rather than queens. At about the time that she starts laying male eggs, she switches off this signal. The goal of the queen through the spring and early summer has been to build up a big workforce. Now, her aim is to use this workforce to produce as many new daughter queens and sons as she can, in the hope of ensuring that she leaves descendants in the following year (I should stress that I am not implying that the queen has actually thought this all through).

This is all very well for the queen: she is equally related to her sons and to her daughters (50 per cent), so she is content to produce both. But rearing brothers is not such an attractive proposition for her daughter workers. Since their brothers have no father, they are equivalent to half-brothers, and so they have only 25 per cent of their sisters' genes. If the nest must produce males, the workers would rather rear their own sons (to which they are 50 per cent related) than their brothers. Although the worker bees are physically unable to mate, they have perfectly functional ovaries, and can lay unfertilised eggs – which will develop into sons.

This conflict over who produces the male offspring leads to chaos. It takes a few days for the workers to detect that there are male grubs in the nest, at which point they start trying to lay their own eggs. The queen cannot tolerate this treachery, and so she sets about eating all of the workers' eggs as quickly as they are laid, thereby consuming her own grandchildren. The queen would much prefer the nest to rear her sons, to whom she is 50 per cent related, rather than her grandsons, to whom she is 25 per cent related. If she catches one of her daughters in the act of laying eggs she will attack and bite her repeatedly; being larger and stronger than her worker daughters she wins the fight easily enough, and then consumes the eggs. However, she is heavily outnumbered. She may win the first fight, and the second and third, but her nest may contain hundreds of daughters, and she cannot bully them all. Now the daughters retaliate, eating the queen's eggs – their baby brothers – and anarchy ensues.

Until recently it would have been difficult to work out who won these battles, but these days it is straightforward to use genetic markers to identify the mothers of males in a nest. Steph has done this with some of her buff-tail nests, spending hours in the lab genotyping thousands of male bees, and she has found that the vast majority of the males are sons of the queen. Other researchers have looked at other bumblebee species and generally found that workers manage to produce at most 10 per cent of males. It seems that, despite being heavily outnumbered, the queen is able to hang on to power for long enough to ensure her own reproductive success.* Nevertheless, the running battles between the queen and her many workers take their toll, and her condition deteriorates – her wings become frayed and her fur thins as she accumulates

* One little-known species, *Bombus wilmattae*, which lives in the mountains of Mexico and Guatemala, has recently proved to be the exception, with about 80 per cent of males produced by workers, but as yet we do not know why.

injuries. From this stage on the nest's days are numbered, for no new workers are being produced, and with some of the existing workers battling over laying eggs in the nest, the incoming food supply dwindles. Sometimes the queen is even killed by her daughters, whilst at other times the food supply runs out and the remaining workers simply wander off, leaving their listless mother to expire amongst the ruins of her nest.

This inexorable process is not as sad as it seems. From the point of view of the queen and the genes that constructed her, even if she is finally killed by the daughters that she patiently reared, her life will have been a success if her genes persist in new queens, by now safely hibernating underground, or as sperm stored inside such queens.

CHAPTER TEN

Cuckoo Bumblebees

The Cuckoo comes in April
She kills a Queen in May
She enslaves her brood
To gather up food
And in July she dies away.

Anon.

Most people are familiar with the cuckoo's nefarious habits. Much as it may seem anthropomorphic to impose human values on a bird, it is hard not to find the cuckoo's habit of laying eggs in the nests of others rather underhand and distasteful, particularly since the baby cuckoo goes on to slaughter its nest-mates, pushing the helpless chicks out.

Far less well known, however, is the fact that many other birds, such as moorhens and various ducks, will also routinely sneak their own eggs into the nests of other birds. They behave much like a cuckoo, waiting until the nest owner is absent before swiftly depositing one or two eggs. The unsuspecting owner returns and subsequently spends time and energy looking after extra offspring not its own. This sneaky strategy's advantage is that it enables the interloper to produce more offspring than it could look after by itself.

Remarkable recent work by Carlos Lopez-Vaamonde at the Institute of Zoology in Regent's Park, London, has shown that bumblebee workers do something very similar. Lopez-Vaamonde was using DNA markers to measure how many males within the nests of buff-tailed bumblebees were produced by workers versus their queen. These nests were housed in a laboratory high up in a building, but the bees were allowed to forage freely through tubes connecting them to the outside world. He found that workers produced only 2.2 per cent of all males within their nest, with the queen producing 95.7 per cent of males. The really interesting result was that 2.1 per cent of males in these nests were not genetically related to either the queen or her daughters, but were the sons of workers from other nests. As most of the egg-laying workers in the experiment were from other experimental nests, it was easy enough to deduce that bees might have become confused and accidentally flown down the wrong tube and hence ended up in the wrong nest. After all, the nests in the experiment were all very close together, and the entrance tubes probably all looked very similar. So far, so good, but what was really remarkable was that some of the workers laying eggs in these nests had come from wild nests somewhere out in Regent's Park or in the gardens nearby. That these worker bees had somehow found their way up to the top of a building and into the experimental nests could surely not have happened by chance.

So what were these bees doing? Remember that, in her own nest, it is not in a worker's interest to lay (male) eggs so long as her mother is laying female eggs, since she is more closely related to her sisters than to her sons. This argument assumes, reasonably, that the nest has finite resources (i.e. food) and cannot rear unlimited numbers of both. But if a worker can get into a nest of entirely unrelated bees, she should not care that any eggs she lays might be reared in place of the queen's offspring. Any reproduction she can get away with in this context is a bonus, increasing the genes she passes on to the next generation. We do not yet know how common this is in more natural situations, but a recent study of the Japanese bumblebee, *Bombus deuteronymus*, found unrelated workers in three of eleven wild nests studied, with these workers producing 19 per cent of the males from these three colonies, so it is clearly not just confined to buff-tailed bumblebees.

A related tactic is also exhibited by bumblebee queens. Some queens emerge from hibernation later than others, even within the same species. Perhaps they choose a particularly shady spot to hibernate, or manage to burrow deeper into the earth where it is cooler. Or perhaps they are just naturally late risers. Whatever the reason, by the time they emerge from hibernation many of the best nest sites have already been taken. One can imagine one of these late queens, repeatedly exploring promising-looking holes in the ground, and each time finding another queen already in residence. As each day passes without her starting her own nest, the season slips away. It takes time to build up a large nest and produce lots of daughter queens and sons, so if she starts her nest too late it is unlikely to be very successful.

In these circumstances, it is common for the queen to attack. If she can kill the resident queen, then she can claim control of the nest site and also take over the resident queen's brood. You may wonder why she should wish to look after another queen's offspring. In most organisms this would be a very silly strategy, but not so in social insects. The brood are destined to become workers, and they will work just as hard for their adoptive parent as they would for their mother, not being able to tell the difference. A similar strategy is used by Australian choughs (birds of the crow family), which live in family groups with just one breeding pair and lots of younger helpers. These groups will readily kidnap half-grown birds from other groups, incorporating them into their team of helpers and thus improving their chances of rearing more offspring.

In the case of bumblebees, by killing the resident queen the intruder is saving herself all the hassle of looking after a colony in its very early stages, albeit at the cost of having to engage in a fight to the death against an individual of similar size and strength. We do not know how common such nest usurpation is, but there are accounts of bumblebee nests being excavated to find as many as twenty dead queens inside, and one live one. Whether all of these dead queens were failed usurpers, or whether the nest had been successively taken over by twenty different queens in succession is impossible to say (at least without DNA fingerprinting the queens and workers, which has not yet been done).

Once a nest has adult workers they should help their queen to repel or kill a usurper, since if their mother dies they are doomed to slavery. For this reason it seems likely that usurping is harder the larger the nest, but on the other hand the prize to be won also becomes greater.

Such usurpation can take place between different bumblebee species, but generally only between closely related species. Buff-tailed bumblebees, for instance, will often try to invade nests of white-tailed bumblebees, but rarely the other way round since the white-tailed bumblebee queens tend to emerge a little earlier than buff-tails. Similarly, in Arctic North America, *Bombus hyperboreus* frequently usurps *Bombus polaris* – and because the Arctic season is so short, the usurping queen does not rear any workers of her own, but instead only new queens and males.

One group of bumblebees, known as cuckoo bumblebees, have become specialists in this tactic, entirely giving up their social lifestyle in favour of life as specialist assassins. There are six species within the UK, all belonging to the same genus as the 'true' bumblebees, *Bombus*. This means that they all have a common ancestor, and would once have all had a similar life cycle, probably similar to most 'true' bumblebees today. But at some point in their evolutionary past, the ancestor of the cuckoo bumblebees evolved down a different route. It is easy to imagine how it happened, and it presumably began as opportunistic usurping, with one late-emerging species often trying to usurp queens of a related, earlier-emerging species. If the likelihood of success in founding a nest by the conventional route was significantly lower than the odds of successfully usurping a queen from an existing nest, then over time the usurping species may have specialised. So doing, they opened up the possibility of evolving physical characteristics to make usurping easier. As a result, these cuckoo bumblebees tend to be larger and have a thicker external skeleton than the 'true' bumblebees, which presumably makes it harder for the resident queen or her workers to sting them to death. It is certainly harder to push a pin through the thorax of a cuckoo bumblebee compared to a 'true' bumblebee.* Cuckoos can also be recognised by the lack of pollen baskets in the females; they don't need them since they don't do any foraging for the nest.

We don't know for sure how cuckoo bees find bumblebee nests to attack, but it must surely be by smell. When keeping artificially reared nests in boxes outdoors it is common to see cuckoo bees flying and walking around searching for an entrance, so they can clearly tell that there is a nest close by. Having found the entrance, the cuckoo bee barges past any workers that get in her way and attacks the resident queen. The queen will usually fight to the death, and with the aid of her workers she may sometimes succeed in killing the intruder. Should she be killed, however, the cuckoo bee will take her place. Occasionally the resident queen will even acquiesce rather than fighting to the death, become subservient to the cuckoo bee and behave like one of the workers. In either case the cuckoo bee will lay eggs, and the bumblebee workers care

* This is not something I do for fun; every entomologist needs a reference collection of pinned bees to refer to when identifying tricky specimens.

for them as they would their own. The cuckoo may also eat any eggs or young larvae in the nest, but tends to leave older larvae to develop into workers which will help care for her own offspring. Moreover, cuckoo bees do not produce their own workers, so the female is not, strictly speaking, a queen. All of her eggs develop into fertile offspring, either males or females. Having taken over a nest, the cuckoo bee queen will continue to lay eggs until the workforce she has coerced into her service begins to die off. There is no supply of further workers, so once taken over by a cuckoo bee, the nest will not last for long. But it will usually survive long enough to produce more cuckoo bees, so continuing the cycle.

The inherited workers continue to work for their new mistress presumably because they have few other options. It is said that they often try to lay eggs, but the cuckoo will chase and bite workers that she finds attempting to do so (just as their mother would have done if she were alive), in a largely successful attempt to keep order. It would be interesting to see whether these enslaved workers are more prone to drifting off to other nests of their own species to try laying eggs there. Presumably this would be in their interests since otherwise they will spend the rest of their days rearing offspring of an entirely different species.

Although cuckoo bumblebees all have a common ancestor, there are now thirty or so species in the world, each specialised to some degree on a particular host. The commonest cuckoo bee in England is usually the southern cuckoo, which targets buff-tailed bumblebees (it presumably being no coincidence that buff-tails are the commonest 'true' bumblebee species). In most respects the life cycle of cuckoo bumblebees is rather similar to that of their hosts. Mating occurs in mid- to late summer, and only the females will hibernate. Males tend to be much more common than females, and can be very abundant – sometimes the commonest bumblebees to be seen – when feeding sluggishly on flower heads of thistles, knapweeds and bramble. Interestingly, the cuckoo species often have a very similar colouration to their hosts – the hill cuckoo, for example, is black with a red tail, and is superficially very similar to its host species the red-tailed bumblebee. Some years ago, it occurred to me that cuckoo females might also mimic the smell of their host; if they did, it would be less likely that the queen and her workers would sound the alarm and mount their defences. All bees are coated in an oily mix of hydrocarbons, the same compounds which make up the smelly footprints on flowers. The precise mix differs between bee species and also probably differs a little between members of different nests within a species, enabling workers to distinguish nest mates from non-nest mates. I started collecting any female cuckoo bees that I could find and storing them in sealed vials in a deep freeze, with a view to analysing them when I had examples of each species. Unfortunately for me, some cuckoos are rather rare and this progressed slowly. Before I could collect enough to do anything useful, I noticed a new paper by Steve Martin at the University of Sheffield in which he had investigated this in some detail, and shown quite convincingly that cuckoo females do indeed have a smell which closely matches that of their host. My samples are still in the freezer.

Although they mimic their hosts in both colour and smell, the disguise of cuckoo bees is clearly not perfect, and they are often attacked. Sometimes I have seen them take refuge in the depths of the comb of the nest, where few workers ever venture. This might well allow them to improve their disguise by covering themselves in the oily hydrocarbons of their hosts, for usually after a day or two in hiding they venture out and assassinate the queen without her workers coming to her aid.

It is easy to think badly of cuckoo bees – I have met people who have been distressed to discover that a bumblebee nest in their garden has been taken over, and have even heard it suggested that we might somehow try to cull cuckoo bees in order to conserve their hosts. This attitude, although understandable, is as nonsensical as condemning a lion for eating a gazelle. Nature is red in tooth and claw, and is so much the richer for it. How sad would it be if we did not hear the sound of (avian) cuckoos in late spring? As we shall see in the next chapter, in addition to cuckoo bees, bumblebees are attacked by a huge range of predators and parasites, all part of a natural community which has existed and co-evolved over millennia. So long as there is enough natural habitat, bumblebees can support this rich diversity of life. Of course the flip side of this is that if we allow bumblebees to disappear, then we will also lose many other fascinating but less well-known creatures besides.

CHAPTER ELEVEN

Bee Enemies

A person who is too nice an observer of the business of the crowd, like one who is too curious in observing the labor of bees, will often be stung for his curiosity.

Alexander Pope

One of the more obvious features of bumblebees is that most are attractively coloured, with bright bands of yellow, and red or white bottoms. They are brightly coloured for a reason: they have a sting. Or at least the females do – it evolved from the egg-laying tube – and they use it to defend the nest against invaders, such as cuckoo bees, or predators. Bumblebees are generally good-natured creatures and almost never use their sting when away from the nest, preferring to fly away if disturbed when foraging. Even in their nest, many species are not very aggressive; I have dug up nests of the early bumblebee without needing any protective gear, the adult bees simply clustering around their brood and buzzing nervously rather than launching an attack. Buff-tails and tree bumblebees are a little more feisty, and their nests are best left alone – on rare occasions I have even been chased by particularly aggressive workers. When really riled, they will bite and jab with their sting at the same time, and are not easily deterred. I hate to admit this, but on more than one occasion I have accidentally pulled the head off a bee when trying to remove one which had fastened its jaws on to my clothing.

It is a common misconception that insects die after they sting. This applies only to honeybees, which have strongly barbed stings which lodge in the flesh of their victim. The bee cannot then escape, and so continues to pump venom until it is swatted – even then the sting and contracting venom sac often remain in place. On the other hand the stings of bumblebees – and for that matter wasps – are not barbed, so they do not get stuck in their victim, and the stinging bee does not die. Hence a bumblebee can, in theory, sting you over and over again until she has run out of venom, so it is fortunate that they generally choose not to do so.

Stings are obviously a very effective defence, but it's even better not to have to use them. That is where colour comes in; as an advert to warn potential predators that this particular prey is armed and dangerous. Many insects which have stings (such as wasps) or are poisonous (such as cinnabar moth caterpillars) have yellow and black stripes, a common signal aimed at predators such as birds that hunt primarily by sight rather than smell. The idea is simple – if birds can't tell which insects are harmful and which are not, they will just attack indiscriminately. By the time they discover that the insect they are trying to eat has a sting or tastes awful, the insect itself may well have been badly damaged. Far better, then, for both bird and insect if the insect broadcasts the fact that it is not good to eat.

The more common the signal, the quicker predators will learn it. This is probably why lots of very different insects use similar signals – yellow and black stripes, or black and red spots (e.g. ladybirds, burnet moths). This leads to one of the great sources of frustration for anyone interested in bumblebees – they are rather hard to identify because many different species use exactly the same colour pattern and so appear very similar. Many species of bumblebee have black and yellow stripes with a whitish tail. Some are black with a red tail. Often, species with near-identical colouration may be quite distantly related, but natural selection has encouraged them to appear as similar as possible. Thus the red-tailed bumblebee and the red-shanked carder bumblebee both have the same colours; the only obvious difference is that the latter has reddish hairs on the pollen basket on its hind legs (from which it gets its common name), while the former has black hairs.

Harmful insects sending out a clear warning is all well and good, but it is a system which is open to cheating, for many insects that don't have any such defences have copied the warning colours of genuinely noxious or dangerous insects, in the hope of fooling predators into avoiding them. Hoverflies provide a good example of a harmless mimic – many of them have yellow and black stripes, similar to wasps or bumblebees, yet they do not have a sting and are perfectly palatable to eat (I cannot confirm this from personal experience). This makes life pretty confusing for the predator. How does it know which signals are genuine and which ones are fakes? If the harmless mimics become too common relative to the models, predators will learn to ignore the warning signal.

Mimicry of this sort can even take place within species. Male bumblebees tend to be broadly similar in colour to their sisters, but they have no sting. Their only protection is presumably the fact that they look like their sisters, and that they appear after a spring and early summer in which the only bees on the wing have been females (with stings). If caught, male bumblebees jab their bottom at their attacker as if trying to sting, presumably in a last-ditch bluff. In fact the males of most species are fairly easy to distinguish to the human eye, particularly since many have tufts of yellow hair on their faces. Once you become experienced in bumblebee identification, you can amaze and amuse your friends by catching male bees with your bare hands. No need to tell them that male bees don't have stings (at least not until later); instead, tell them that you have a natural way with bumblebees, that they trust you. Of course this takes a little confidence; best to practise without a crowd as you are likely to appear quite foolish if you get it wrong – not at all the heroic look you were going for. Bumblebee venom is similarly painful to that of a honeybee, i.e. an initial squeal and some hopping about is hard to suppress. It contains a whole range of chemical compounds, including histamines, which have been 'designed' by evolution to hurt as much as possible.

It has puzzled me for many years that a handful of bumblebee species do not bother to have warning colouration at all. The common carder bumblebee is, for instance, a scruffy brown with a few tufts of black, yet as well-defended as any other bumblebee. Many years ago I became convinced that they must have warning stripes in ultraviolet (which birds and bees can see but we cannot), but after several weeks spent working out how to take pictures in ultraviolet it emerged that they are just as dull in this spectrum as in the one we can see. Interestingly, the tails of white-tailed bumblebees *are* UV-reflective, presumably enhancing their signal to birds.

All of this said, it is rather rare to see predators eating bumblebees. In North America, robber flies and beewolves will take smaller bumblebees. Robber flies are splendidly ugly, powerful flies with humped backs, which grab their insect prey in mid-air. Beewolves are wasps which catch bees on flowers – they sting them to paralyse them, and then feed them to their grubs. In Europe, beewolves seem to mainly eat honeybees; I have watched them for hours returning with prey to their burrows on the sandy banks of the River Charente in France and have never once seen one bring back a bumblebee. European robber flies are also mostly too small to tackle a bumblebee. There are some birds further south in Europe that are well known for eating bees – bee-eaters being the most obvious. The species of bee-eater found in southern Europe is an exotic, colourful bird, with a russet cap, blue wings and a yellow throat. They are very agile and fast fliers, snapping up bees on the wing with their sharp, down-curved beak, then deftly nipping off the sting. I have also seen a flock of them perched on fence wires next to honeybee hives in the Sinai Desert,* picking off every bee as it tried to leave the hive. I'm sure they must take bumblebees too, but bumblebees tend to be rare in southern Europe because it is too warm for them.

Shrikes also eat bumblebees. These rather splendid birds have the gruesome habit of impaling their surplus prey on the spines of thorn bushes as a snack for later. They feed on a range of big insects such as bumblebees and grasshoppers, and also frogs and small mammals. Shrikes are moderately common in parts of southern Europe but are very rare in Britain, so here at least it seems unlikely that they pose much of a threat to bumblebees.

Not so long ago I was of the opinion that bumblebees had a pretty easy time in Britain, there seemingly being few predators able to attack them. Smaller workers sometimes get caught in spiderwebs, and are occasionally trapped by crab spiders,† but

* Beekeeping is not a traditional activity in the Sinai because for most of the year there are very few flowers so honeybees do not thrive. However, in recent years some enterprising individuals have taken to keeping honeybees which they feed with sugar syrup. The bees recycle this into honey, which the beekeepers harvest and then sell to unsuspecting tourists as 'traditional' Bedouin honey.

† Crab spiders are so named for their rather crab-like appearance, being flattened with their legs protruding at the side, but their most striking characteristic is their colour. These are sit-and-wait predators that perch on flowers, waiting for insects to visit. They are coloured to match the petals, many species being white or bright yellow. My logical mind tells me that they are quite beautiful, but they still give me the willies.

otherwise they seem pretty safe. Or so I thought, until the spring of 2008, when I was sent a pile of dead queen bumblebees, enough to fill a large strawberry punnet, by a Mrs Barbara Baker who lives in the west of Scotland. She had picked the bees off her lawn, from beneath an overhanging sallow tree. There was a whole range of species, including many bilberry bumblebees which are rather rare in most of Britain. Each bee had been carefully dissected, with the top of its thorax removed and all of the flight muscles scooped out, leaving the thorax looking rather like a hard-boiled egg after one has spooned out the contents. Barbara hadn't seen the culprit, but a bird seemed likely. Sallow catkins are a popular source of food for bumblebee queens in spring and clearly something was attacking the bees in the tree.

By coincidence, in July of the same year I received a letter from Anne-Marie Smout, a keen naturalist who is heavily involved in biological recording in central Scotland. Anne-Marie and her husband Chris had just come back from holiday in Denmark, where they had been staying with friends. The friends' garden had some large lime trees. Buff-tailed and white-tailed bumblebees love the flowers of lime trees, although there is something in the nectar which seems to make them dopey and even sometimes to kill them. One morning when having breakfast on the lawn, Anne-Marie noticed many dead bumblebees under the trees. On closer inspection she found that each had been attacked in the same way – the back of the head had been opened up, and the brain scooped out. The abdomen had also been opened and hollowed out. Intrigued and slightly horrified, Anne-Marie and Chris set out to discover the culprit by hiding themselves out of sight and watching the tree with binoculars. They didn't have to wait more than a few minutes. A whole family of great tits emerged, a pair of adults with their young, and they continued their banquet. Being apparently drunk on the lime nectar, the bees were easy prey to the birds, which had presumably learned how to avoid the stings and peck open the bees' bodies.

I wrote a short article on this for *Buzzword*, the newsletter of the Bumblebee Conservation Trust, and this provoked further records of great tits eating bumblebees. In particular places, the birds seem to have developed different techniques. Some birds seem to open up the thorax, others nip off the tip of the abdomen, still others go for the head, or different combinations of the three. Tits copy one another, so presumably one of the adults in the family that Anne-Marie observed had discovered how to eat bumblebees, and the rest of the family copied the technique. It is reminiscent of the spread of milk-bottle-opening behaviour in blue tits in the 1960s. When I was a child, it was normal to have milk delivered to the doorstep. It arrived in glass bottles with foil lids. Every doorstep would have a tile or piece of wood to hand, which the milkman would balance on top of the bottles. If he did not, within minutes a blue tit would appear, peck open the foil, and drink the cream from the top of the milk. I must admit to feeling slightly sad that this rather cute behaviour has all but died out, as most people now buy their milk at the supermarket, or have the low-fat variety delivered, which has no cream on top.

Luckily for the bees, cases of great tits attacking them in this way seem to be very scattered and sporadic, and usually associated with a large flowering tree that is attracting lots of bees. However, this isn't the end of this particular story, as we shall see.

Although foraging bumblebees may not have too many predators, bumblebee nests are a different matter. A large nest contains stocks of honey and pollen, and many tasty grubs and pupae, as well as the adult bees: a very valuable food resource to any creature able to overcome the defences of the bees. Steph O'Connor, my bumblebee sniffer-dog handler, has spent the last few years studying the predators and parasites that attack bumblebee nests. She has become remarkably good at finding bumblebee nests, considerably exceeding the abilities of poor Toby. Her project, funded by the Leverhulme Trust, was to find nests (with or without the aid of the dog), and then to follow their progress for the rest of the season, recording what attacked them, and whether they survived to produce new queens and males.

She and I investigated various camera systems, and in the end adopted one which had been developed by the RSPB for monitoring birds' nests. The system has a small camera mounted on a post and trained on the bumblebee nest. The camera is connected to a large, buried Tupperware box in which resides the recording equipment and battery. Once installed the whole set-up is very inconspicuous, the camera being painted olive green for camouflage and nothing else being visible. The software records continuously, but immediately deletes the information if no movement is detected around the nest entrance. This greatly reduces the amount of data storage necessary, so that data needs downloading (and batteries renewing) only every few days. The camera is sensitive to infrared and has its own built-in infrared light so that it can detect movement through the night. In 2010 and 2011, Steph used this set-up to film thirty-six bumblebee nests from when she found them until they expired. The footage consisted of every 'event' during the remaining life of the nest – rather tedious to watch since most of the events were bees entering and leaving, or grass and leaves blowing in the wind. But every now and then the cameras revealed something exciting and new. No one has ever watched bumblebee nests in this way before.

It turned out that the most common visitors were mice, voles and shrews. Some nests had a steady traffic of small mammals going in and out, but we still don't know exactly what they were doing. Charles Darwin was interested in the relationship between mice and bumblebee numbers. In his most famous publication,

On the Origin of Species, he suggested that red-clover pollination (which is exclusively carried out by bumblebees) was better in the vicinity of villages, for villages have cats and cats eat field mice, which would otherwise eat the bumblebees:

> 'The number of humble-bees in any district depends in a great degree on the number of field-mice, which destroy their combs and nests; and Mr H. Newman, who has long attended to the habits of humble-bees, believes that "more than two thirds of them are thus destroyed all over England." Now the number of mice is largely dependent, as every one knows, on the number of cats; and Mr Newman says, "Near villages and small towns I have found the nests of humble-bees more numerous than elsewhere, which I attribute to the number of cats that destroy the mice." Hence it is quite credible that the presence of a feline animal in large numbers in a district might determine, through the intervention first of mice and then of bees, the frequency of certain flowers in that district!'

Thomas Huxley, a great friend and supporter of Darwin, took this apparent logic even further by arguing that the power of the British Navy was attributable to spinsters, who keep cats, which eat mice, which therefore don't eat bumblebees, enabling the bumblebees to pollinate clover, which is fed to cattle, from which salted beef was derived, which was the staple food of the British Navy. In fact there is little evidence to back any of this up, although it makes a lovely story. It is possible to turn the logic entirely on its head, because bumblebees regularly use the abandoned nests of small mammals such as mice to create their own nests. We don't yet know the net effect of this interaction, although an educated guess might suggest that, for bumblebees, there is an optimum number of small rodents, sufficient to provide abandoned nests but not so many that most bumblebee colonies are then eaten. Despite the frequency of small mammal traffic in and out of Steph's nests, she found no evidence that nests visited by many small mammals were more likely to die, suggesting that the mammals were not eating large numbers of brood. For all we know they may simply be sharing an entrance hole with the bees, for the tunnels to which such holes lead are usually variously branched and can go deep underground. Until we can find a way to get the cameras into the nest itself – which is a possibility using a flexible endoscope – we will not know for certain what is going on underground.

As I have mentioned, badgers are among the best-known predators of bumblebee nests; they are seemingly impervious to stings and consume the entire nest, bees and all, if they can dig it up. When I have placed artificially reared bumblebee nests in the field I have often had them torn to shreds by badgers, for they are powerful animals, and Perthshire farmers who use commercial bumblebee boxes for raspberry pollination often suffer from similar problems. Luckily for Steph's nests, which were mostly in and around the campus of Stirling University, there appear to be no badgers here. She did have other larger mammals such as squirrels investigate the nests, and hedgehogs regularly came by and snuffled around the entrances, but they were too big to get in and not good enough at digging to burrow down. On one occasion the camera recorded a cow's hoof standing on the entrance hole and squashing it, but the bees soon repaired the damage. Foxes and weasels have also been claimed to attack bumblebee nests, but we have no idea how often or how serious these predators are, and Steph recorded neither visiting her nests. In North America, skunks are said to be partial to bumblebees, whilst in Iceland, escaped mink are thought to be important predators. Of course these days there are plenty of escaped mink in Britain, but whether they are eating bumblebees is not known.

Perhaps Steph's most exciting discovery was that great tits do not confine themselves to plucking bumblebees from flowering trees. A number of Steph's nests were regularly visited by great tits, which would sit by the entrance, picking off foragers as they came and went, and returning day after day. In general it seemed to be the biggest nests that were attacked, probably simply because these had more traffic going in and out and so were easier for the great tits to detect. Other observations of note were bees carrying grubs from their nest and dumping them outside – we have no idea why they do this, but they may have been sickly, or possibly unwanted male grubs. Steph also recorded red-tailed bumblebees entering a buff-tailed bumblebee nest. Again, we don't know why, but they may be trying to steal honey. This work is in its early days, but we hope to learn much more about the relationships between bumblebees and their nest predators as it goes on.

As well as trying to record predators, Steph has looked into the other less obvious creatures that attack her bumblebee nests. From every nest that she has filmed she has taken weekly samples of faeces from the workers. Just as a doctor might inspect a stool sample to help diagnose a human illness, Steph screened these samples for gut diseases. Bumblebees can suffer from a range of nasty gut infections including both protozoans and fungi, some of which give them chronic diarrhoea, and these infections are usually evident when the faeces are inspected under a microscope. Some of the disease-causing organisms even swim around energetically like tiny tadpoles. Extracting faeces from bumblebees is a fiddly business. If they are contained within a plastic pot they sometimes defecate quite quickly, perhaps because the poor things are scared. Gentle agitation tends to speed the process up but some bees absolutely refuse – maybe they simply don't need to go. If they do produce a sample, it has to be sucked up swiftly using a tiny glass tube – bee faeces are very liquid – or else the bees tend to stand in it and smear it all over the place.

Steph found that many of these diseases were common – for example almost every single nest she studied was infected by a gut parasite known as *Crithidia bombi* by the end of the season. *Crithidia* is a protozoan, related to the organism which causes sleeping sickness in humans and livestock in Africa. It is thought to spread rapidly through direct contact between nest mates, and spreads between nests via flowers, which get smeared with the disease by infected workers. *Crithidia* doesn't do too much harm to individual workers so long as they have plenty of food, but if food is in short supply then the disease becomes more serious and can kill them. If young queens become infected they are generally less likely to survive through the winter and manage to found a new nest in spring, but young queens appear to have especially strong immune systems so they rarely get infected.

Once Steph's nests died off, she attempted to dig them up to find out more about what had happened to them. This can be hard and frustrating work. I once spent a whole day trying to dig out an unwanted buff-tailed bumblebee nest in someone's garden. The tunnel turned erratically underground, eventually going under a tarmac road so that I could follow it no further, leaving me with a trench about 10 feet long to fill back in. Many of Steph's nest tunnels led amongst the roots of large trees, forcing her to abandon excavation. For perhaps half of them, she managed to get down to the nest itself. There was usually rather little left of the original structure. One of the major bumblebee nest predators seems to be the bumblebee wax moth, a small, nondescript cream-coloured moth that lays batches of eggs in bumblebee nests. When the eggs hatch, the caterpillars spin tough silken tubes that protect them from the bees, and tunnel through the nest, indiscriminately consuming wax, pollen, grubs and pupae. In heavy infestations, the nest is completely destroyed, the bees seemingly unable to do anything to fight back. These wax moths also seem to be particularly prevalent in bumblebee nests in gardens, perhaps simply because bumblebee nests are common in gardens so there is plenty for them to eat; 80 per cent of buff-tailed nests that I placed out in gardens in Hampshire were attacked by wax moths, many of them completely obliterated. The majority of Steph's natural nests were also attacked. Once they have finished demolishing the nest, the caterpillars huddle together to hibernate in a conspicuous white mass of tightly woven silk tunnels, which is often the most obvious thing remaining.

Where nests have not been destroyed by wax moths, there is often much to see. The body of the old queen is usually there, sometimes with those of other queens who attempted to usurp her. Corpses of workers litter the nest, scattered amongst the wax nectar-storage pots and empty pupal cocoons. Queen cocoons are larger than those of workers or males, and so the number of them gives an indication of the number of new queens produced by the nest. Steph found that many nests produced few or no queens, but just a few very large nests managed to each produce more than one hundred.

Usually there are also lots of other insects in old bumblebee nests, so called commensal organisms which have no particular effect on their hosts but live alongside them, scavenging on bee faeces and other scraps; these include various flies, moths and beetles. There is often an assortment of maggots – fly larvae – of varying sizes wriggling around. *Volucella bombylans*, for instance, is a beautiful large furry hoverfly that mimics bumblebees; it occurs in various forms, which resemble different bumblebee species. Sladen observed that this fly is sometimes attacked and killed by the worker bees when trying to enter the nest, but that if this happens the fly immediately lays all of her eggs in her death throes, which the worker bees rarely notice. When the eggs hatch, the tiny maggots scurry away into the bottom of the nest where they can scavenge undisturbed. The beetle *Antherophagus nigricornis* finds its way into bumblebee nests by climbing on to a flower. It waits there for a bumblebee to arrive, then grabs hold with its mandibles. Being about 5 millimetres long, it is inevitably noticed by the bumblebee but she usually cannot shake it off. Once she gives up and returns to her nest, the beetle drops off and thereafter resides within the nest.

Aside from all of these insects, bumblebee nests almost invariably also contain a multitude of tiny mites. Mites are relatives of ticks and slightly more distantly of spiders; they have eight legs, and the ones found in bumblebee nests are usually tan-coloured. As with the maggots, they are scavengers within the nest. Maggots turn into adult flies, which can readily fly off to locate a new bumblebee nest each year, but mites cannot fly at any stage of their life, which poses a problem for them. Bumblebee nests die off each year, so that any mite left after the bees have gone will soon starve. To solve this problem, some mites climb on to new queens before they leave their natal nest, and then hibernate with them, hoping to colonise the new nest that they will attempt to found in the spring. These mites can be very common, so that many of the bumblebee queens that emerge in spring have clusters of mites living in the crevices between the thorax and abdomen, behind their head and between the bases of their legs. They look revolting, but most do not actually feed on the bee – they are just hitching a lift. Some smaller mites live permanently amongst the bees' fur. One very small species actually lives on the back of the larger mites, themselves on the bees, bringing to mind the phrase, 'A flea hath smaller fleas that on him prey; and these have smaller fleas to bite 'em, and so proceed ad infinitum' (Jonathan Swift, 1733). These mites probably do very little harm to their hosts, although they must weigh them down a little. Only one species of mite lives exclusively within bumblebees, inside their trachea (breathing tubes), sucking on their blood, and thus presumably they do their host no good at all.

Aside from the mites that hitchhike on queen bees, there is one other bumblebee parasite that specialises in attacking queens. This fascinating but repulsive beast is the nematode worm, *Sphaerularia bombi*, a distant relative of the pin worms and hook worms that infect humans. The young female worm burrows into a queen bee while she is hibernating in the soil, and takes up residence within her abdomen. Peculiarly, the worm everts her ovaries so that they are outside her body – these then grow enormously by absorbing nutrients from the host, to around 2 centimetres in length. They resemble a knobbly white sausage, dwarfing the rest of the worm's body which is just visible to the naked eye as a tiny thread attached at the side. Somehow, the worm alters the behaviour of the poor host queen when she emerges from hibernation. Instead of trying to found a nest, she sits around listlessly, occasionally feeding, until May or June and then returns to her hibernation site where she wanders around on the soil surface, defecating. Her faeces are full of a multitude of tiny new worms; a single adult female can produce 100,000 or so. These burrow into the soil where they live for a few weeks, and mate. When the next generation of queen bees burrows into the soil to hibernate, the worms will be waiting for them.

Many other organisms have evolved to parasitise bumblebees. Wasps of the family *Braconidae* and flies of the family *Conopidae* lay their eggs in adult bees when they are feeding on flowers, with their larvae then consuming the host from the inside. When infected with the latter, worker bees sleep outside the nest at night where it is cooler, slowing the development of their parasite and so delaying their inevitable death. Flies of the family *Sarcophagidae* fly into bumblebee nests and give birth directly to maggots (there is no egg stage), which feed upon bumblebee pupae, slowly eating them from the outside. Velvet ants (*Mutilidae*), not ants at all but ant-like wingless furry wasps, will also walk into bumblebee nests and lay their eggs on bee pupae, which once again are then consumed.

Overall, bumblebee nests support a vast diversity of fascinating and sometimes rather stomach-churning creatures, many of which have been little studied, and about which we know only scraps. Clearly the importance of bumblebees in supporting biodiversity goes far beyond their role as pollinators of wild flowers.

As yet, we do not know how many bees die as a result of predators and parasites, but Steph's work is beginning to give us some idea. The nests she has filmed suffer from an alarming mortality rate. Many die within a week or two of being discovered, despite Steph taking great care to leave them undisturbed. So far as we can tell, this is natural – many, many queens attempt to start nests in spring, but very few succeed in rearing a large healthy nest. Perhaps as many as 50 per cent of all nests die every two weeks. Fortunately those few nests that manage to grow to a large size then produce many queens – making up for the large majority of nests that have failed. Of course this means there is a delicate balance, and any factor that increases mortality even a little could push a bumblebee species into rapid decline.

CHAPTER TWELVE

The Birds and the Bees

Concerning the bees and the flowers
In the fields and the gardens and bowers,
You will note at a glance
That their ways of romance
Haven't any resemblance to ours.

Anon.

Since moving to the University of Stirling I have taken to running up a local hill, named Dumyat (pronounced Dum-i-at), every Wednesday at lunchtime. The university campus is lovely, the ugly 1970s buildings hidden amongst trees around a beautiful loch. Just to the south-east rises a steep volcanic plug clothed in woodland, on which has been built the dramatic Gothic tower of the Wallace Monument. The Ochil Hills rise directly from the north-east end of the campus, and form a sharp escarpment running eastwards into Fife, marking the edge of the central Lowlands and the beginning of the Highlands beyond. Dumyat is the westernmost of the Ochils, rising only 400 metres or so, but with spectacular views into the Highlands to the north, to Edinburgh in the east and to Ben Lomond in the west.

I moved to Stirling in 2006, and in the late summer of that year I noticed something a little odd at the top of Dumyat. I was gasping to regain my breath, slumped against the cairn that marks the summit, when I noticed several bumblebees buzzing about. Closer inspection revealed that they were male white-tails. They were not feeding on flowers – indeed, there are no flowers anywhere near the top of Dumyat, just rocks and sheep-cropped grass. It is a fairly bleak and windswept spot – the bees were being buffeted about, but were stubbornly flying into the wind to maintain their position, whilst zigzagging around as if looking for something.

A week or two later I took my family for a walk up to the base of the Wallace Monument, a short but steep climb through the woods, which emerges into a grassy clearing at the top. There were a few flowers about, mostly creeping thistles, and they were covered in male forest cuckoo bumblebees, sometimes two or three to a flower. I was intrigued as to what these gangs of male bees were doing hanging around at the tops of hills. I enlisted the help of two Stirling undergraduates, Jill Young and Liz Sangster, persuading them that they should spend their summer walking up and down a range of Scottish hills while counting the numbers of male and female bumblebees, to see if this was a general phenomenon. They duly did so, and it seems that it is – every hill they climbed had unusual numbers of males at the top, and comparatively few on the slopes and at the bottom.

The explanation for the predilection of male bumblebees for hilltops is most likely that they are engaging in a behaviour appropriately known as 'hilltopping', something which has been studied in detail by John Alcock of Arizona State University, now retired but one of the world's experts on insect mating behaviour. He has devoted his life to unravelling the mysteries of how insects find and choose their mates, doing much of his work in the exciting-sounding but presumably rather inhospitable Sonoran Desert. One recurring theme of his work is that males of many species, including such evocatively named beasts as the great purple hairstreak, the pipevine swallowtail and the tarantula-hawk wasp, aggregate on hilltops where they await the arrival of females. Often the males become territorial, choosing a particular perch and vigorously chasing away other males who encroach on their space. Competition for perches at the very top of the hill, which seems to be the prime location, can be fierce. Females of most insects need to mate only once or perhaps a few times in their life, while males try to mate as many times as they can. In hilltopping species, females that wish to find a mate simply head uphill, confident that when they get to the top there will be plenty of amorous suitors awaiting them. This also provides them with an opportunity to be choosy; they can quickly survey and compare a large number of males, and offer themselves to whichever one of them has the most desirable characteristics. Or, if they are in a hurry, they might simply mate with the male right at the top, on the assumption that he must be the fittest and strongest since he has managed to acquire and hang on to the best spot. With luck their sons will inherit these characteristics too, and will themselves become king of the hill. Once mated, the females quickly depart, usually to lay their eggs, perhaps returning to find a second mate a few days later when they have depleted their sperm stores.

In bumblebees, the males do not appear to be territorial when gathered on hilltops. They simply hang around, often in amicable groups, drinking nectar when it is available, and presumably attempting to mate with females whenever they appear. The parallels with human behaviour are irresistible, particularly since male bees do little or no work in the nest, leaving all of that to the females. Hilltops seemingly represent the bumblebee equivalent of a nightclub or singles bar, somewhere to check out and chat up members of the opposite sex.

Unfortunately there is a rather crucial aspect in which the analogy falls down. To my knowledge, no virgin queen bumblebees have ever been observed heading for the high ground. Indeed, this is a recurring theme in studies of bumblebee mating behaviour. Prior to Jill and Liz's work on hilltopping, many other odd behaviours have been described in male bumblebees and interpreted as mechanisms for locating a mate, but virtually none of them has ever been observed to succeed. Actual mating in bumblebees is seldom observed, and usually only once copulation has already started – the process of successful courtship is almost never seen in the wild. In fact, most bumblebee males never do mate. For reasons we do not yet understand, bumblebee nests seem to produce many more males than new queens; on average about seven males for every queen. Since queens generally mate only once in their entire lives, this means that six out of seven males will never manage to mate. As it is their sole function in life, I find this rather sad.

The most intriguing supposed mate-location behaviour used by bumblebees is often called 'patrolling'. Charles Darwin was deeply interested in this, and employed his many children as convenient and cheap labour to help him study patrolling bumblebees in his garden in Kent. He had noticed streams of male garden bumblebees passing along the hedges and ditches, generally all heading in the same direction and passing by every few seconds.

> 'I could only follow them [the male bees] along this ditch by making several of my children crawl in, and lie on their tummies, but in this way I was able to track the bees for about twenty-five yards.'

I guess the children should have counted themselves lucky that he didn't send them up to sweep out the chimneys too. Darwin noted that the bees would pause every few yards at what he described as buzzing places, before continuing along their route. To make it easier to see the bees as they flew, he equipped his children with sugar-sprinklers filled with flour, with which they sprinkled the bees as they passed, giving them a ghostly white appearance.

> 'The routes remain the same for a considerable time, and the buzzing places are fixed within an inch. I was able to prove this by stationing five or six of my children each close to a buzzing place, and telling the one farthest away to shout out "here is a bee" as soon as one was buzzing around. The others followed this up, so that the same cry of "here is a bee" was passed on from child to child without interruption until the bees reached the buzzing place where I myself was standing.'

Fascinatingly, Darwin found garden bumblebee males patrolling more or less the same route in successive years, despite the fact that male bees do not survive from one year to the next.

Since Darwin, many other researchers have described patrolling in a range of other bumblebee species, including white-tails, buff-tails and red-tails. Scientists in Scandinavia seem to have taken a particular interest in this subject in the 1970s and 1980s, and thanks to their efforts, we now understand much more about this behaviour. Different species seem to prefer to patrol routes at different heights, which perhaps helps them to avoid accidentally mating with members of other species. For example, red-tailed bumblebees prefer routes across the treetops, while garden bumblebees patrol routes low to the ground, presumably explaining why it is garden bumblebees that are most often seen patrolling by humans. The males mark out the route in the morning, pausing every 5 or 10 metres to paint a pheromone on to a twig or leaf. The pheromones are produced in the labial glands inside their head, and painted on to the chosen object using a beard of bristles on their mandibles. Each species has a distinctive pheromone blend, which once again presumably helps to avoid confusion.

These pheromone marks become the 'buzzing places' described by Darwin, where later in the day the bees simply pause and hover briefly. Each male patrolling a circuit will paste a different object at each buzzing place, and so will hover in a slightly different place, but all follow the same broad route. Routes are typically 200 or 300 metres long and male bees are swift fliers, so each bee can repeat the circuit every few minutes.

This elaborate and remarkable behaviour has only one conceivable purpose: to attract a mate. We presume that the pheromones attract virgin queens, and that the males circulate as fast as they can to maximise their chances of being the first to find any queen that comes along. However, queens are seldom seen to show any interest. My best guess is that when one does come along, she is rapidly found by a male and they quickly disappear off into the undergrowth to mate, so one would have to be very lucky to see it happen. I have tried collecting the pheromone from several dozen male red-tailed bumblebees by squashing their heads in a suitable solvent,* then pasting the concentrated pheromone on to fence posts on the edge of Yew Tree Hill nature reserve in Hampshire. This area teems with red-tails so there should have been new queens around, but none came to my pheromone lures. Perhaps if I had painted the pheromone on to the tops of trees, where the males normally patrol, this would have worked.

It remains a mystery how males come together and agree the route, whilst the persistence of these routes over successive years adds an additional puzzle. Perhaps traces of the pheromones linger

* You are probably thinking that this doesn't sound very nice. Of course the males were humanely killed before their heads were squashed. I should also remind you that there are always many more males than are needed to mate with the small number of queens, so this would not have done any harm to the local population.

from year to year, or perhaps some physical attributes of the landscape – for example the shape, size or orientation of a particular hedge – are inherently suitable for a patrolling route in some way that is beyond our ken and so attract males year on year.

Hilltopping and patrolling are not the only strategies used by male bumblebees in apparent attempts to find mates. Males of a few species seek out nests from which virgin queens are emerging, and then hang around the entrance. How they find the nests, we do not know – I would guess it is probably by smell. When studying bumblebees in the Outer Hebrides, one of my PhD students, Ben Darvill, once observed a swarm of male moss carder bumblebees outside a nest. He wondered whether they were hanging around their own nest, perhaps with a view to making incestuous advances towards their sisters, and so took DNA samples of the males and some of the female workers. It is relatively straightforward to test whether bees are siblings using genetic markers. These were not, and neither were most of the males brothers of one another – it seemed that the males had come from a variety of other nests.

This is not to say that male bumblebees are averse to incest, as another of my PhD students, Penelope Whitehorn, has found. Penelope, a pretty and terribly well-spoken girl from Oxfordshire, spent much of her PhD encouraging brother and sister bumblebees to mate and then studying the consequences. She placed virgin queen buff-tailed bumblebees in cages with either a brother or an unrelated male, and found that the male bees showed no scruples whatsoever, vigorously attempting to copulate with the queen regardless of whether she was their sister or not. On the other hand, the queens were a little more restrained; although they often did allow their brothers to mate with them, on average they held out a bit longer when confined with a brother than when with an unrelated suitor. This all makes perfect sense. Males of most animals are very unfussy about whom they mate with, because they can mate many times and it costs them little – just some sperm. On the other hand, females often mate rarely and invest a huge amount in rearing their young, so they have a strong incentive to ensure that their offspring are sired by a healthy, attractive and unrelated father. Hence they tend to be much more picky. Breeding with close relatives is particularly risky since it can lead to genetic problems in the offspring, so most animals have mechanisms to avoid it if they possibly can.

Males of most common UK bumblebees do not swarm outside nests, but those of our newest species, the tree bumblebee, do so with great enthusiasm. This species invaded Britain from France in 2001 – quite by chance I was the first person to catch one here, when out hunting for bees in the New Forest. It is a very pretty and distinctive little bee; chestnut brown and black with a white bottom. The name derives from their penchant for nesting in holes in trees, and they also readily take over tit boxes. So far as we know this was a natural invasion, and they seem to be doing no harm. The press showed some interest in the arrival of this new species and dubbed it 'Le Bee', even speculating that it might have a distinctive French buzz, which sadly it does not.* Tree bumblebee

* My name was mentioned in a few newspaper articles about this new bee. I subsequently received a storm of emails and letters describing creatures which the authors presumed were tree bumblebees. The summer of 2001 happened to be a good year for hummingbird hawkmoths in the UK, and I received many accounts of a hovering insect with a very long tongue which was surely a tree bumblebee. One lady described how she had been terrorised at night by an insect 'with huge staring eyes' which had attempted to get in through her bedroom window. I guess this was also a large moth of some sort – their eyes can reflect light like a cat's – but she was sure that it was a tree bumblebee.

males swarm in their hundreds around nests, and since the nests are often in garden tit boxes, this can cause considerable alarm. Of course male bees cannot sting so they pose no danger, but it is not always easy to convince alarmed homeowners not to call in Rentokil.

There is one mate-location strategy used by bumblebees which I have not yet seen, largely because none of our UK bees do it. Only a handful of species do, and the males of all of them have distinctively large, bulging eyes. Three such boggle-eyed beasts can be found in North America – the Nevada bumblebee, the brown-belted bumblebee and the red-belted bumblebee – and these have been studied in some detail by Kevin O'Neill of Montana State University. The males select prominent perches in the early morning, and paste them with pheromones in the same way as do the patrolling species. As the day warms up, they sit upon their perch and wait for females to heave into view. Despite their large eyes, they don't seem able to distinguish queen bees from other large insects, so they dart out indiscriminately at any passing creature of roughly the right size. If it is a queen of their species they then try to grab her in flight, and subject her to an enthusiastic sexual assault in mid-air. O'Neill found that the pairs often flew away locked together, and he couldn't be sure whether they mated or not. Some couples crashed to the ground and these he was able to watch, but in most cases the female was able to shake or kick off the male and escape by scurrying into dense vegetation. The males are fiercely aggressive towards one another, attempting to drive rivals from nearby perches or fighting to oust each other from the best perches, and these fights can result in serious injuries. Despite this the strongest males can hang on for a long time – O'Neill found one male on the same perch for twenty-six consecutive days.

Most of the research on bumblebee pheromones and mating behaviour has focused on males, but some fascinating and slightly gruesome studies carried out by John Free in the 1970s showed that virgin queens also produce a sex pheromone. Free was a great expert on the behaviour of honeybees and bumblebees who worked at Rothamsted Experimental Station (now Rothamsted Research) in Harpenden. He published dozens of papers on bees over forty-five years from the 1950s to the 1990s, and he has a particular interest in pheromones. Free found that if he tethered live, virgin bumblebee queens near the 'buzzing places' used by patrolling males, they were quickly located and courted. If he tethered old, mated queens, workers or males in the same places, the passing males very sensibly showed no interest. He went on to discover that males would happily attempt to mate with freshly killed virgin queens. Free reasonably assumed that the males could distinguish virgin queens from mated queens by smell; after all, they look the same. To determine which part of the queen produced this smell, he experimented with removing body parts from dead virgins. Males were undeterred by the removal of most body parts, but seemed unwilling to mate with decapitated queens. You might put this down to squeamishness, but if the headless bodies were smeared with chemicals squeezed from the head then they once more became attractive and the males resumed their amorous advances. Males were not completely indiscriminate; when presented with crude dummies smeared with the odour of virgin queens they showed brief interest, but quickly departed without attempting to mate, indicating that smell alone is not enough to keep them interested for long.

We should not judge males harshly for their fairly relaxed attitude as to what constitutes an eligible mate; with only a one-in-seven chance of mating at all in their life, they cannot afford to be especially fussy. Because mating in bumblebees is rarely seen, it tends to provoke considerable excitement when it is. I have been sent several photographs of male bees clinging enthusiastically to the back of a queen of entirely the wrong species, although in these circumstances it seems that they rarely persuade her to cooperate. Sometimes several males of an assortment of species will be pictured pushing and shoving one another for pole position, a tangle of legs and fur on the back of a disgruntled queen.

Although I have never observed successful courtship and copulation in the field, bees do mate readily enough in the lab. I recently played host to a Bangladeshi researcher named Ruhul Amin who was particularly interested in courtship and mating in bumblebees, and he spent many happy hours in a windowless room in the basement at Stirling watching buff-tailed bumblebees mate. He found that older males took longer to mate than young ones, but that they had better endurance, mating for longer. He offered some lucky males a chance to mate with a second virgin immediately after they had finished with the first, and found once again that older males took longer to perform in these circumstances, although when they eventually got going they mated for longer. He also found that bigger males mated more swiftly but had less endurance.

Amin found that during courtship the males invariably climbed on to the back of the queen. He observed that they held the thorax either side of her head with their forelegs, while stimulating her nether regions with their hind legs. The queens are much larger than the males, so the males are unable to force them to mate if they are not interested. Fascinatingly, when Amin confined a queen with two males, he found that on average the successful male was not the youngest or the biggest, but the one with the longest fore- and hind legs. We presume that longer limbs make them better able to hang on to and simultaneously stimulate the female.

As with most insects, the male bee has a powerful pair of claspers at the tip of his abdomen, with which he grips the female, and a tube-like endophallus which is inserted into the female and carries the sperm. Once his claspers are locked on, he lets go of the female with his legs and 'lies back', legs folded in front of him as if meditating, with the only point of contact between himself and the female being the genitalia. Mating typically lasts for about half an hour. Sperm is transferred in the first minute or two, but after that the male pumps a sticky glue into the queen, forming a 'mating plug', which prevents her from doing it again, and explains why queen bumblebees generally mate only once.

As you may have gathered, in general I think that bumblebees are among the most fascinating of creatures, and their mating habits are certainly intriguing and enigmatic, but in this one area I have to concede that their cousins the honeybees have them trumped. Unlike bumblebees, honeybee queens mate a dozen or so times, all during a single nuptial flight made early in life. Each young queen releases a pheromone which is enormously attractive to the males – known as drones – who eagerly swarm after her. They mate in mid-air, the male grasping the female's abdomen with his claspers and then explosively squirting sperm into her. This explosion produces an audible pop and ruptures the male's genitalia, usually causing fatal damage so that he falls to the ground and quickly expires, leaving behind fragments of his genitalia still attached to the queen. She continues her flight, and is swiftly mounted by a second male who scrapes off the remains of his predecessor before repeating the process. The new queen continues until she has stored enough sperm for a lifetime of egg production, at which point she returns to the nest, leaving behind a trail of dead lovers.

CHAPTER THIRTEEN

Does Size Matter?

I beg a million pardons. Abuse me to any degree, but forgive me: it is all an illusion about the bees. I do so hope you have not wasted any time for my stupid blunder – I hate myself, I hate clover and I hate bees.

Charles Darwin,
letter to John Lubbock, 1862

In 2001 I took on a new PhD student named James Peat. James had been an undergraduate at Southampton, and my tutee. From a slightly posh family in Dorset, James was a bit of a charmer and a ladies' man, often sporting amusing facial hair. I'd always found him to be very entertaining, with a quirky, lateral view of the world, and I thought he might be just what my research group needed. Among other things, he is the only man I've ever known who insisted on making all of his own trousers.

I decided to let James loose on a topic which had intrigued me for many years. Bumblebees differ from their close relative the honeybee in that the workers are enormously variable in size. The largest workers are roughly ten times heavier than the smallest, yet all live alongside one another in a single nest. Most insects vary little in size: one adult peacock butterfly is much the same size as another; and the same can be said of grasshoppers, dragonflies or stag beetles, for example. Even the relatives of bumblebees

– wasps, stingless bees and honeybees, – tend to be more or less uniform in size. The only group of insects which exhibits something similar to bumblebees are the ants. In some ant species, most notably leafcutters from the rainforests of South America, the workers vary enormously in size and this relates to the jobs that they perform. The biggest workers are monsters, with huge scimitar-shaped jaws suspended from large heads packed with muscle. Their job is to defend the nest against large vertebrate predators such as anteaters or insectivorous birds and lizards, and they can deliver a ferocious bite. The medium-sized workers forage for food, forming trails through the forest and up the trunks of trees to the foliage, where they snip semicircles of leaf to carry back to the nest. Collectively, they can strip a huge rainforest tree of leaves in a single night, so each night they set out in a different direction.

Tiny workers tend to the queen and brood, and also chew up the leaves brought in to the nest to feed to their fungus garden, in chambers deep underground. Leafcutter ants cannot digest leaves directly, so they have evolved a remarkable symbiosis with a species of fungus found only in their nests. The white, fluffy fungus turns unpalatable leaves into digestible fungal fronds, which are harvested by the tiny workers and fed to the rest of the ants in the nest. Some of the tiny workers also hitchhike on the backs of the foragers collecting leaves, and are thought to protect them from parasitoid wasps and flies, which would otherwise inject their eggs into the foragers.*

It clearly makes sense for different workers to have bodies suited to their jobs; the tiny workers would be hopelessly ineffective if

* Anyone with even a passing interest in ants should consider reading the marvellous *Journey to the Ants* by Bert Hölldobler and the legendary biologist E. O. Wilson.

called upon to attack an anteater, while the giant soldiers would not be small or nimble enough to care for the fungus garden. Does a similar line of reasoning explain why bumblebee workers vary so much in size?

The first step towards finding out the answer to this question was to determine if different-sized bees do different jobs. In fact it had long been suspected that the smallest bees tended to stay in the nest, and the bigger ones went foraging for food, but this had not previously been quantified. To do so, we simply reared up thirty buff-tailed bumblebee nests, then placed them out in the field and opened the doors to their boxes. These boxes consisted of an outer cardboard shell with a removable inner plastic box, and they were fitted with two doors, one which allowed bees both in and out, and the other with a valve which only allowed them in. We left the nests to settle down and forage naturally for four weeks, after which we returned in the middle of the day, when foragers were likely to be out doing their thing, and shut the doors. We then replaced the inner plastic box with an identical but empty one, and opened only the door which allowed bees in. Foragers returned home for the next few hours, loaded with food, but found that they had entered an empty nest from which there was no escape. James then had the tedious job of using Vernier callipers to measure the size – the width of the thorax – of all the bees from each nest, both those that were inside the nest when we sealed the door, and those that had been out foraging; about 4,500 bees in total, roughly one-third of which had been foraging.

This was not a perfect experiment; I am sure you will have realised that many foraging bees might have happened to be in the nest at the time we shut the doors. Nonetheless it showed very clearly that the very smallest workers were all in the nest, suggesting that they rarely if ever go foraging – in fact, unless you look inside a bumblebee nest you will never see these diminutive creatures. Medium-sized workers seemed to include a mix of nest bees and foragers, while the very largest workers were almost all foragers, just as had been suspected.

This is all very well, but why are foragers bigger than nest bees? It is easy to understand why ants which are to defend the nest must be large, but less easy to see why carrying food necessitates being large. You may well be thinking, 'Surely a large bee can carry more?' It would be surprising if they could not, but it isn't as simple as that. Big workers presumably require more food to rear in the first place. Let's suppose, for the sake of argument, that it takes the same amount of food to rear one big worker or two small ones of half the weight. Further, let's suppose that the big worker could carry twice as much food. On each foraging trip, the two small workers would bring back the same weight of food between them as the larger bee. However, they would be back sooner as they would only have to visit half as many flowers to gain a full load – all else being equal – so they would be able to make more trips per day, and should be able to bring back more food in total. It would only be worthwhile to produce large foragers if they were somehow a lot better at gathering food than small ones, or if they had some other significant advantage of which we were unaware.

James set out to discover how efficient bees of different size were at gathering food. His experimental set-up was simple. We rigged up a buff-tailed bumblebee nest in my lab with a clear plastic tube attached to the nest door. When the door was opened, bees could walk down the tube, through a clear plastic chamber constructed over the pan of an electric balance, along a further tube, and out through the window to the outside world. We could individually identify every bee since we had carefully glued a numbered plastic disc to the thorax of each. This is an enormously fiddly job. The glue takes a minute or two to set, during which time the bee tries to dislodge the disc by hooking a foreleg over its back. As often as not it succeeds, and the process has to be repeated. Not infrequently the bee pushes the sticky disc forwards on to its head, effectively covering its eyes, and then blunders around bumping into things. I once tried using quick-drying superglue instead, but after ending up with a bee with its foot stuck to its back I quickly gave up.

As each bee left the nest over the pan of the balance, its number and weight were manually recorded. Initially, bees tended to fly across the chamber rather than walking over the pan, but this problem was solved by covering the chamber with transparent red plastic film; bees can't see red light very well so the chamber appeared dark, and bees won't fly in the dark, so they were forced to walk. Once they left the tube protruding from the lab window, the bees invariably hovered around for a minute or two, presumably trying to memorise the location, before disappearing off into the suburban gardens of Southampton. It was very exciting watching the first few bees depart, and we sat about eagerly awaiting their return.

After about half an hour, the first bee appeared back at the window, with large balls of pollen on its legs. Unfortunately it didn't seem able to find the end of the plastic tube, so it flew backwards and forwards along the long row of glass windows. It was soon joined by others, and quickly a cloud of confused bees built up outside the window. It became apparent that there was a flaw in our plan. The lab was on the third floor of a very sizeable and ugly concrete building, with seven floors in all. Viewed from the outside, there were seven rows of identical windows, one above the other, and each row was perhaps 50 metres long. All looked much the same, and the little plastic tube protruding from the corner of one of them was exceedingly inconspicuous. Our poor bees knew their nest was there somewhere, but couldn't find the correct window. Before long the frustrated bees started coming in any open window, often a floor or two above or below my own, invading offices, biochemistry labs and so on. Since the bees all had numbered discs attached it didn't take anyone long to realise who was to blame, and we soon found ourselves bombarded with phone calls and having to run around the building with nets and pots to catch our strays. We shut the nest door so that no more bees could go out, put back as many as we could recover, and sat down to discuss what to do. Eventually we decided on attaching a large orange funnel to the nest tube, providing an eye-catching target for returning bees. To make life even easier for them, Ben Darvill, then an undergraduate student, somehow acquired a traffic cone and gaffer-taped it to the outside of the building just below the correct window. How he got it up there I never found out, since it was too large for him to have put it out through the window. When I left Southampton five years later it was still there.

The funnel and traffic cone served their purpose and bees began foraging from the nest, running out through the plastic tubes and returning laden with pollen or nectar. Once this was all working, James and a team of undergraduate helpers set about recording and weighing every bee that left or came back into the nest. Someone had to be by the nest at all times during daylight hours, so they set up a rota between them. The nest lasted for about one month, after which we replaced it with a second, and eventually a third, so the team, between them, spent three months sitting at the lab window with notebook in hand. The difference between the outgoing and incoming weights of each bee allowed us to calculate how much food they had gathered on their trip, and since we also knew how long this had taken them, we were able to calculate their efficiency in terms of forage gathered per minute. Some bees were notably more dynamic than others, sprinting down the tubes, returning within a few minutes with full loads, then immediately setting off again. Others ambled along the tube, stopping occasionally, sometimes getting to the exit funnel and then turning back as if they'd changed their mind and decided they didn't really fancy going foraging after all. The keen bees were enormously impressive. Bumblebees carry pollen on their legs and nectar in their honey stomach, a chamber which, when full, almost fills their abdomen. Some of them brought back up to 150 milligrams of food, close to their own body weight, having gathered it in twenty minutes or less, and they did so many times during the day.

Bees got better at foraging with experience. On their first trip from the nest, many came back weighing less than when they set out, presumably because they had flown about burning energy and failed to find any food. Over time their success rate increased, and on average it took about thirty or so trips before they reached peak efficiency. Presumably during this time they were learning about how to navigate through the landscape and experimenting with different flowers to see which were most rewarding. On their first foraging trips bees tended to collect nectar, usually returning with no signs of pollen on their legs, but as they got older and more experienced they turned to collecting more pollen. I suspect that this is because drinking nectar is much easier than gathering pollen, for the latter requires the bee to brush pollen from the anthers with her hairy legs, then comb the grains from her body into her pollen baskets, using a little nectar to glue them together. An experienced bee makes this look easy, but it must take quite a bit of practice.

A few largish bees never went foraging but sat for most of their time just inside the door to the colony, or just outside in the tube. We dubbed these 'guard bees', since honeybee colonies have individuals that sit in the nest entrance and try to prevent intruders, but in truth we don't really know if these bees were actually doing any guarding. They never did anything when foragers came past them – often literally walking over them – perhaps because all the bees coming in were legitimate members of the nest. If a cuckoo bee had attempted to get in, or a worker from another nest hoping to lay a few eggs, they might have sprung into action.

It turned out the bigger foragers were, on average, more efficient at gathering food, bringing back more per unit time. The smallest of the bees that left the nest rarely returned with a net profit and so contributed very little to colony growth. This presumably explains why foragers tend to be large, but in doing so it raises another question: why are bigger bees better at foraging? I discussed this with James, and our first idea was that it might relate to keeping warm. To fly, a bee needs to keep her core temperature above about 30°C, and bigger bees ought to be better able to do this since they have a smaller surface area to volume ratio, but they might be prone to overheating on hot days. Bees working in the nest don't need to worry about keeping warm since they don't have to fly and in any case the nest itself is kept very cosy by insulation and the combined activities of the bees inside. If a bee becomes too cold when foraging she will be grounded, unable to return home, and at great danger of being eaten by a predator, so this might explain why small bees avoid foraging. James tried testing whether the smaller of the bees that went foraging were affected more badly by cold weather, but this didn't seem to be the case. He also went through the data set carefully to see whether smaller foragers tended to go out only on warmer days, but this didn't seem to be true either.

James and I didn't really get to the bottom of what made bigger bees better at foraging, but part of the answer has since been provided by a German researcher named Johannes Spaethe, working in the lab of Lars Chittka at Queen Mary University, London. He found that larger workers have much more acute vision, in part simply because their eyes are much bigger. This is likely to be enormously important when searching for particular types of flowers in the landscape, or when memorising and distinguishing between landmarks for navigation, but of no value whatsoever to a bee operating in the dark confines of the nest. He also found that having larger eyes enables bigger bees to fly in darker conditions, presumably helping them to get home if they find that it is starting to get dark when they are out collecting food. Spaethe went on to study the sensitivity of antennae to odours, and discovered that larger workers also have a better sense of smell. This would benefit them greatly when trying to locate or identify flowers by smell. Larger bees also tend to have bigger brains and so might have enhanced learning ability; this too would help in the complex job of traversing the landscape in search of food, although it has not yet been studied. Finally, larger bees might just be more likely to survive when out and about in a dangerous world. I have commonly seen bumblebees tear themselves from spiderwebs, and I imagine that this would be much harder for a small bee. Despite their large size, the life expectancy of foragers – at two to three weeks, roughly the same as a First World War pilot – is much lower than that of nest bees, which often survive for months. This difference might be much worse if the foragers weren't larger.

Interestingly, honeybees have a different strategy in this respect. Rather than divide up jobs according to size (workers are all the same size), honeybees divide up jobs by age. The younger bees stay in the nest, looking after the brood, while the older bees do the foraging. It seems likely that older bees are given the most hazardous job because they have less residual value to the colony; to take an extreme example, a very old bee is likely to die any day anyway so why not send her on a perilous mission, whereas a young bee has many weeks of work left in her, and it would be a greater loss to the colony if she were to die. There is a possible parallel with humans, although you might find this far-fetched. In the Hadza, a hunter-gatherer tribe of Tanzania, older, post-menstrual women gather a disproportionate amount of food, spending longer out in the savannah picking berries and digging for tubers than their daughters and granddaughters. In evolutionary terms, it may be a wise strategy for post-reproductive women to spare their children such arduous and potentially dangerous tasks and take the risks themselves, since they are unlikely to live too much longer and they are not able to reproduce again themselves. In evolutionary terms, granny's genes would rather she were eaten by a lion than that her grandchildren were eaten. I should add that I am not suggesting this is a conscious strategy, just as I do not suggest that worker bees consciously choose to help rear their sisters because they are aware of patterns of relatedness.*

Let us return to size variation in bumblebees. The question as to why foraging bumblebees tend to be large can be turned on its head by asking, why are nest bees small? It seems likely that the small bees are more nimble in the confined space of the nest, so that it benefits the colony to have small workers for within-nest tasks and big ones for food-gathering. If this were so, nests with workers of a range of sizes ought to fare better than nests with only large or only small workers. Such nests don't naturally exist, but James and I set out to create them by moving bees around. In many social insects, it is not possible to move workers between nests; for example, if one places a worker ant in a different nest she is usually swiftly attacked and killed. Social insects can usually detect the odour of non-nest mates, and are not welcoming to visitors. However, I had noticed that bumblebees from commercial nest boxes often enter the wrong nest when several are placed near each other, and they seem to be accepted. I presume that this is because they have

* My students often object when I suggest that human behaviour might in part be explicable in evolutionary terms, but many human emotions – which drive behaviours – such as romantic or parental love, or jealousy, have a clear purpose and are readily explained in terms of natural selection.

been reared in identical conditions, and so perhaps they all smell much the same. Thus I thought that it might be possible to juggle workers between nests to create all-big or all-small colonies.

I experimented with a giant home-made pooter. Pooters were invented as a device for picking up small insects without harming them; such creatures are all too easy to squash with forceps or fingers. Ordinarily, a pooter consists of a small glass bottle with a cork in the top. The cork has two holes in it, through which two long, flexible, transparent tubes are pushed. Importantly, the end of one of these tubes, where it protrudes below the cork inside the bottle, is covered with fine mesh or netting. To use a pooter, one tube is placed in the mouth – this must be the one with the mesh attached – and the other tube is pointed at the insect. A swift suck on the former usually results in the insect shooting up the latter in a rush of air and ending up, unscathed, in the bottle. It is best to avoid doing this with ants as they squirt out formic acid when frightened which is not kind to the lungs and results in a lot of coughing, but otherwise this is an excellent device for handling tiny beasts. It had occurred to me that a super-sized pooter based on a large jam jar and tubes large enough to suck up bumblebees might be just the thing for efficiently moving bumblebees between nests. I constructed one, and tried it out.*

* I once spent a morning with a class of eight-year-old Spanish children, attempting to imbue them with an enthusiasm for insects. We walked out on to the nearby flowery hillside where I armed each of them with a butterfly net and a pooter. I was pleased to see that the children were in tremendous high spirits, laughing uproariously as I tried to explain how to use the pooter in my very poor Spanish, when their embarrassed teacher whispered in my ear that *puta* is the slang word for prostitute, going on to explain that it is also widely used as the equivalent of the F-word in English.

I had a nest in our darkroom which was equipped with red lighting so that the bees would not fly when I removed the lid. Several workers were scurrying around nervously on top of the nest, aware of the disturbance. When handling bumblebee nests it is important not to breathe on them as this causes the bees to become very agitated, presumably because they think they are under attack from a large mammal such as a badger. I carefully pointed one tube at a bee, and sucked hard on the other. The bee clung on, confused by the sudden draught. I gave the mightiest suck I could manage, and was starting to go blue in the face when the bee suddenly rocketed up the tube and into the jar. Pleased with my success, I paused to get my breath back, then tried a second bee. Unfortunately I had forgotten the all-important netting on the tube going to my mouth. As I sucked, the bee that was already in the jar shot up the tube and into my mouth. Before I could spit it out it stung me, causing my lower lip to swiftly swell to a ridiculous size. As you might imagine, James and my other students made fun of me mercilessly for the rest of the day.

My giant pooter never really caught on, but nonetheless James did manage to juggle bees between nests to create four different experimental treatments: nests with thirty large workers, nests with thirty small workers, nests with about sixty small workers of the equivalent total weight as thirty large workers, and nests with thirty workers of assorted sizes. These he placed out on the university campus to see how they fared. We predicted that the nests with only thirty small bees would be poor at foraging and so grow very little. The nest with thirty large bees ought to be good at collecting food but poor at looking after their young. Comparing nests with sixty small bees versus thirty bees of twice the size ought to reveal whether two small bees are better than one big one. Finally, we predicted that the nests with a mix of large and small bees, as occur naturally, would perform best of all.

When the bees had had three weeks of looking after themselves, James gathered in the nests and counted how many young they had reared. Our predictions were wrong. The nests with only big workers reared the most offspring, performing better than nests with a natural mix or with the equivalent weight of small worker bees. Taken at face value, this suggests that bumblebees have got it wrong. They shouldn't bother rearing any small workers, but should just concentrate on producing big ones, even if this means having fewer of them. This may be true – no organism is perfectly adapted to its environment – but I suspect that the results might be misleading. The weather during James's experiment was awful, with near-constant rain, and none of the nests were thriving. Perhaps larger bees are better at coping with such adverse conditions. It may serve the queen's interests to have mainly small workers in the nest, for they would be easier for her to dominate and prevent from laying their own, male, eggs. Perhaps mixing bees from different nests is hopelessly unnatural and causes strife within the nest, disrupting its normal functioning. We still don't fully understand the roles that bees of different sizes play in bumblebee nests.

There is one final twist to this tale. When out and about watching bees it soon becomes obvious that different species of bumblebee visit different species of flower. For example, almost the only visitors to foxgloves are garden bumblebees, plus the occasional common carder bumblebee, as only these species have a long enough tongue to reach the nectar. Similar but more subtle differences in floral preferences can be found *within* species, according to variation in size, not least because this relates to tongue length. Although foraging bees tend to be larger than nest bees, there is a lot of size variation within the foragers. In buff-tails, the tongues of the smallest foragers are just 4 millimetres long, while those of the largest are nearly twice this length.* James caught and measured the size of hundreds of wild, foraging buff-tails in and around Southampton, and found that the average size and tongue length varied greatly depending on which flower he caught them on. White clover, bramble and oilseed rape attracted the smallest bees, greater knapweed and field beans the largest. The latter two flowers have deep corollas. He also recorded how long it took bees of different sizes to extract the nectar from particular flowers. Lo and behold, large bees with long tongues were quicker at getting nectar from deep flowers than smaller bees, but the converse was true on shallow flowers, where small bees came into their own.

James noticed that the floral preferences of bees of a particular size were also influenced by the sturdiness of the flower. The biggest workers are pretty chunky beasts, the largest insects likely to land on most flowers. Some flowers simply cannot cope with their weight. Occasionally, a big buff-tailed bumblebee will land on a white clover, at which point the stalk of the flower often gives way and it collapses to the ground. The bee is left struggling to feed while lying on her back under the flower, which looks neither comfortable nor particularly efficient. They usually give up on white clover pretty quickly and go to find something else to feed on. In contrast, other flowers have very sturdy stems: viper's bugloss and knapweed flowers can easily support the weight of large bees, and so provide them with a more attractive proposition.

All of this means that it makes absolute sense for a colony to

* Measuring the length of a bee's tongue is a fiddly business. They don't readily agree to it, so they need to be anaesthetised or cooled in a fridge before the tongue is carefully unfolded and measured with callipers.

produce foragers ranging in size, for this allows it to efficiently exploit a broad range of flowers in the surrounding area: small, short-tongued bees for shallow and weak-stemmed flowers, and bigger, longer-tongued bees for the sturdier and deeper flowers.

You might by now be wondering as to the relevance of the Darwin quotation at the start of this chapter. It was taken from a letter to his friend John Lubbock on 3 September 1862. The context is that the day before, Darwin had written to Lubbock with a request:

> 'I write now in great Haste to beg you to look (though I know how busy you are, but I cannot think of any other naturalist who wd be careful) at any field of common red clover (if such a field is near you) & watch the Hive Bees: probably (if not too late) you will see some sucking at the mouth of the little flowers & some few sucking at the base of the flowers, at holes bitten through the corollas. All that you will see is that the Bees put their Heads deep into the head & rout about. Now if you see this, do for Heaven sake catch me some of each & put in spirits & keep them separate. I am almost certain that they belong to two castes, with long & short probosces. This is so curious a point that it seems worth making out. I cannot hear of a clover field near here. Pray forgive my asking this favour, which I do not for one moment expect you to grant, unless you have clover field near you & can spare 1/2 hour.'

Darwin had been watching honeybees foraging in a field of red clover near Southampton, where he had recently been staying, and had noticed that some were robbing the flowers through holes chewed at the base, while others were feeding from the conventional entrance to the flower. Darwin was familiar with ant species in which castes with different morphology perform different jobs, and he came to the notion that this might also apply to honeybees; he supposed that those feeding conventionally had long tongues, while the short-tongued bees robbed flowers from the side. He then fired off the request to Lubbock to test this idea for him. It seems that after sending this request, Darwin himself must have decided to sample some bees and measure their tongues. He realised that he had sent his friend on a wild goose chase – as we now know, honeybee workers are all of much the same size and tongue length – and so he sent his angst-ridden apology the very next day, finishing with, 'I hate myself, I hate clover and I hate bees.' It seems a little over the top, for so far as we know Lubbock never found time to look into this in the day between the arrival of the two letters. Of course, had Darwin asked his friend to carry out the same exercise with bumblebees, he might very well have found a positive result.

CHAPTER FOURTEEN

Ketchup and Turkish Immigrants

Time is honey.

Bernd Heinrich
(American entomologist and marathon runner)

It is a common misapprehension that there is just one species of bee: they have yellow and black stripes and they sting; they live in wooden boxes, where they are looked after by bearded old men in funny hats and white suits; they pollinate crops and wild flowers; and they produce honey. Of course by now you know that much of this is untrue. There are perhaps 25,000 species of bee in the world. The bees kept by beekeepers are honeybees, and they don't have yellow and black stripes – they are, in fact, largely tan-coloured. Some beekeepers themselves often believe or assume that all crop pollination is carried out by their honeybees, but in reality honeybees are absolutely hopeless at pollinating some crops such as runner beans or tomatoes, whilst they are good at pollinating others such as oilseed rape or kiwi fruit. Depending on their size, shape, behaviour and the length of their tongues, different bee species are suited to pollinating different types of flowers, and some plants are better pollinated by moths, or flies, or beetles.

It has long been known that bumblebees are effective pollinators of many crops and wild flowers. It was for this reason that they were taken to New Zealand (to pollinate red clover), and why the commercial rearing of bumblebees was developed by Dutch scientists in 1988, initially for tomato pollination. Breeding of bumblebees en masse is a tricky business. Back in 1912, Frederick Sladen described how to get queens to build nests in captivity, but he did not have the facilities to mate bumblebees or to hibernate the queens through the winter, so he could only get them to complete part of their life cycle. Since then, dozens of scientists and amateur enthusiasts have played around with rearing bumblebees. Sladen found that imprisoning queens for brief periods in a nest box with food would sometimes induce them to start nesting. He also discovered that putting two queens together seemed to help. One generally became subservient to the other, and the dominant one was likely to start laying eggs. Adding workers of the same or closely related species, or some pupae from another nest, also seemed to encourage queens to lay. Peculiarly, later scientists found that adding young honeybee workers also helped to stimulate the bumblebee queen into action; this technique is still widely used, and it is an odd sight to see a huge bumblebee queen in her nest being attended by relatively tiny and slender honeybees.

Sladen's facilities were limited to what he could build in his father's garden shed, but the widespread availability of equipment for maintaining controlled environmental conditions made life much easier for prospective bumblebee rearers in the second half of the twentieth century. It was discovered that queens seem to nest much more readily if kept at a constant 28°C and high humidity. Mating proved to be fairly easy so long as young queens and males were confined in reasonable numbers in brightly lit cages, and it was discovered that newly mated queens would readily hibernate in loose soil and could then be stored for months in a refrigerator.

By the 1970s enough had been learned to make it possible to

take some bumblebee species through an entire year in captivity. Buff-tails seem to be particularly easy to rear, but other species such as the short-haired bumblebee are much more stubborn and even today very few people have managed to rear so much as a single nest in captivity.

The commercial possibilities of bumblebee rearing were not immediately realised. Most of those who had helped to develop the techniques were scientists who simply wanted nests to study and use in experiments. Rearing bumblebees was still highly labour intensive, and there seems to have been considerable scepticism that artificial rearing of nests could be commercially viable. This changed rather rapidly in 1985, when a Belgian veterinarian and amateur bumblebee enthusiast named Dr Roland De Jonghe found that placing a buff-tailed bumblebee nest in a glasshouse full of tomatoes provided a remarkably effective pollination service. Some crops, such as tomatoes and peppers, require buzz-pollination, the rapid vibration of the flower. The pollen-producing anthers are rather like an inverted pepper pot – they have to be shaken to release the pollen, and bumblebees are particularly adept at this. They grasp the anthers in their mandibles and then use their flight muscles to madly vibrate their body and the flower, at the same time adroitly catching the pollen as it falls out. Up until that time, tomato growers had been hand-pollinating tomatoes – employing teams of labourers with vibrating wands to touch every flower three times a week. The labour costs were enormous, amounting to about €10,000 per hectare per year. At that price, the effort of rearing a few bumblebee nests is comparatively trivial. What is more, De Jonghe found that both the quality and quantity of fruit produced was higher when pollinated by bumblebees. He realised he was on to a gold mine, and began rearing bumblebees for sale.

In 1987 De Jonghe founded the company Biobest, which remains today one of the largest commercial producers of bumblebees. In

1988 they produced enough bumblebees to pollinate just forty hectares of tomatoes. By the following year they were exporting to Holland, France and the UK. Others wanted in on the action; the Dutch company Koppert Biological Systems began rearing bumblebees in 1988, followed by Bunting Brinkman Bees, also Dutch, in 1989. The competition led to improvements in rearing techniques, which are jealously guarded by each of the producers, and drove down the cost of the nests. By 1990, the Canadians were using commercial bumblebees; the next year the USA and Israel joined them. Japan and Morocco soon followed suit. By the end of the millennium, bumblebee pollination had become the industry standard for tomatoes in almost every country in the world (excluding Australia, as we saw earlier).

Today, there are at least thirty factories in the world producing bumblebees, mainly buff-tails. Many of them originate from Turkey – for some reason Turkish buff-tails seem to be particularly amenable to mass production. The European factories produce well in excess of a million nests per year, and there are shipped all over the globe. I am one of very few people who have been inside several of the major rearing facilities, but I had to sign confidentiality agreements to get in, and hence, sadly, I am not at liberty to describe them in any detail. Suffice it to say that the scale of the operations is staggering – imagine vast white rooms the size of football pitches, with tall stacked ranks of bumblebee nests on shelves in row after row stretching into the distance, tended by teams of technicians in white lab coats sweating in the warm, sticky conditions.

From an environmental point of view, the switch to using bumblebees for tomato pollination has one substantial advantage. Growers have to be extremely careful in their use of pesticides, or else they risk killing their bees. This has forced many to turn to natural control agents, such as predatory insects, against the many pests that attack tomatoes – for example parasitic wasps to deal with whitefly – thereby substantially reducing pesticide use and providing healthier food for us to consume. Unfortunately there are also several downsides to commercial bumblebees, some of which will have become apparent from my description of the consequences of the arrival of bumblebees in Tasmania. In the Antipodes, there are serious risks that escaped bumblebees might worsen weed problems, but elsewhere there are other issues which are perhaps even more worrying.

Rearing and distributing bumblebees has a substantial carbon footprint; the factories themselves are vast and require heating and lighting. The nests are housed in disposable boxes made of a plastic inner box surrounded by polystyrene, in turn encased in cardboard. These boxes are often burned by farmers – they are not intended to be recycled. The nests are transported thousands of miles, sometimes across continents, to the places where they are used. None of this is good for the environment.

A second concern relates to the risk of these bees escaping into the wild and competing against or hybridising with native bees. Some of the commercial suppliers have argued that bumblebees can be contained within glasshouses so they won't escape. This argument was put forward by the horticulture industry when lobbying the Australian government to allow the introduction of bumblebees to mainland Australia. Presumably fur farmers in the UK said the same thing before their mink escaped and ate most of our water voles. It is ludicrous to argue that bumblebees will not escape if tens of thousands of nests are imported and distributed across hundreds of farms, even if they are used in glasshouses. Glasshouses have to have vents for warm days, and doors for people to go in and out, and windows get broken, so of course bees will escape. In the UK, imported nests come with advice recommending that they should be destroyed after use by incineration or freezing, but in my experience many farmers throw them on a skip or simply leave them in place until all the bees have gone. Few farmers have sufficient freezer space to fit in the nests, and burning plastic and polystyrene boxes full of live bees is smelly, creates a cloud of toxic fumes, and seems rather cruel – poor thanks for the pollination services the bees have provided.

In Japan they have brought in laws insisting that all glasshouses in which bumblebees are to be used have netted vents and double doors, but since buff-tailed bumblebees have already escaped into the wild in Japan and are thriving there, this seems to be very much a case of shutting the glasshouse door after the furry black-and-yellow 'horse' has bolted. In any case, netting vents greatly reduces their effectiveness, increasing humidity and leading to fungal disease on the crops, so farmers are not keen to go down this route.

Just as in the UK, native Japanese bumblebee species seem to be having a hard time of it, for Japan is a similarly small and crowded island. Some Japanese researchers have suggested that the arrival of non-native buff-tails is adversely affecting some native bumblebees, but it is too early to be sure. The biggest threat may be to *Bombus hypocrita*,* a relative of the buff-tail. It seems that young *Bombus hypocrita* queens are all too easily wooed by the suave foreigners, and readily mate with them despite their being of the wrong species. Such liaisons are disastrous for the queen for she only mates once, and sperm from buff-tailed males are unable to fertilise her eggs, so she is doomed to sterility. Recent studies suggest that 30 per cent of *Bombus hypocrita* queens suffer this fate in areas where buff-tails have become abundant.

Similar problems may be occurring in Britain and Ireland, where

* I couldn't find out the Japanese common name of this bee, but in Korea it goes by the rather musical name of *Sap-po-lo-dwi-yeong-beol*.

we have a distinct subspecies of buff-tailed bumblebee, formally known as *Bombus terrestris audax*, found nowhere else in the world (apart from New Zealand and Tasmania, to which they were introduced). The most obvious difference is the tail colour of the queens – our queen buff-tails have, as you might surmise, buff-coloured tails, although the workers' tails are more or less white. In Europe, both the queens and the workers have white tails. The bees being imported to the United Kingdom, roughly 60,000 nests per year, are from Turkey and Greece and belong to the subspecies *Bombus terrestris dalmatinus*, although some of them may also be from France and Germany, in which case they would be *Bombus terrestris terrestris* – some of the factories are no longer quite sure of the origins of their stock, and to my eye they look much the same. In cages, British buff-tails readily mate with their Continental cousins, be they from Turkey or France, but unlike the situation with *Bombus hypocrita* this results in viable, hybrid offspring. Whether this is happening in the wild in Britain we do not yet know, but it seems likely. The hybrid offspring would not be obvious, since the buffness of the tails of our native queens is quite variable anyway. Hybrids might have buff tails, white tails, or something in between – we don't know, and I very much doubt anyone would notice whichever it was. It may be the case that Continental *Bombus terrestris* have escaped into the wild in the UK and established their own populations. They might compete with our native bees for food and nest sites, or they might not – we simply don't know. If anyone spotted a queen bee in the UK with two yellow stripes and a pure white tail they would assume it was *Bombus lucorum*, the species we call the white-tailed bumblebee.

The only way to get a handle on this issue is to do extensive genetic testing. Lucy Woodall, a postdoctoral researcher in my lab at Stirling, has recently been doing some work on this in her spare time. She has found a genetic marker that appears to distinguish between our native buff-tails and those from the Continent, but she has also detected small numbers of the Continental type in samples taken from the wild in Britain. She has looked at only one gene so far, so we cannot say whether these are hybrids or pure-bred European bees. We also cannot be certain that bees of the European type don't naturally occur in Britain – they might occasionally make it over the Channel. There is a simple way to find out though – by using museum specimens. If the Continental types currently in the UK result from the commercial trade, there shouldn't be any of them among samples of buff-tailed bumblebees caught before 1988. Modern genetic techniques make it pretty easy to get DNA from old, dried specimens, provided the museum staff agree to let us chop a foot off some of the bees in their collections*. With luck we may have a definitive answer soon, although I'll have to find someone to do the work as Lucy has just left Stirling for a job at the Natural History Museum in London, where

* This is an interesting issue. Examining the DNA from old specimens, and comparing it to that of their present-day descendants, can provide exciting insights into how species have evolved and been affected by the environmental changes of the last hundred years or so. At present, obtaining DNA requires the removal of a small part of the body – usually a foot in the case of insects – and some museums have become loath to allow this. Not so long ago the techniques were less sophisticated and a whole leg was needed. Some particularly important specimens are now lacking many of their limbs, which is very sad. The museums argue that genetic techniques advance almost daily, and that anything we can do today could probably be done much better tomorrow, perhaps with less damage to the specimens. I can see their point, but it can be very frustrating when answers are needed now rather than later, and also this argument could be used indefinitely.

she will be studying the DNA of deep-sea worms. My grant applications to fund work on this have thus far been rejected, but I shall keep trying.

Of course if we find that commercial bees are established in the wild in Britain, or that they have hybridised with our native bees, there is not much that we can do about it. It may be that the trade in commercial *Bombus terrestris* has irrevocably muddled the genetics of this species throughout its range, which extends through much of Europe and a chunk of Asia. If it has, then we have lost for ever the local races that once existed. Sad though this might be, I do not think that it is the biggest cause for concern with regard to commercial bumblebees.

My greatest worry relates to disease. The diseases of honeybees have been studied fairly intensively for many years as they have obvious impacts on beekeeping and honey production. Honeybees are known to suffer from a broad range of viral, bacterial, fungal and protozoan diseases, as well as larger parasites such as mites. In contrast, before the 1990s we knew almost nothing about bumblebee diseases until a Swiss zoologist named Paul Schmid-Hempel turned his attention to them. Paul and his students have been prolific in their findings and much of what we now know about bumblebee diseases comes from them, but as I am sure Paul would admit, we have still only begun to scratch the surface. It has become clear that bumblebees suffer from as broad a range of diseases as do honeybees. Many of them are fairly closely related to the honeybee diseases – for example, honeybees are attacked by the protozoan *Nosema apis* while bumblebees are attacked by the sister species *Nosema bombi*. Bumblebee viruses have not yet received serious attention, but it seems that some viruses known from honeybees also turn up in bumblebees. The descriptively named Acute Bee Paralysis Virus and Deformed Wing Virus have both been found in wild bumblebees, and for all we know they may well also infect other bee species and perhaps other insects. At the moment, we know almost nothing about the natural geographic ranges of most of the diseases that infect bumblebees, which bumblebee or other bee species they infect, what harm they do to their hosts, and to what degree there are different strains of diseases in different countries.

Unfortunately, mass-rearing of bumblebees provides a great environment for the multiplication of disease, no matter how carefully it is done. All diseases spread more quickly when their hosts are packed closely together. Taking bees reared at high density and then shipping them thousands of miles to places where the bees themselves and any diseases they might be carrying do not naturally occur is an incredibly risky strategy. In fact, if one wished to spread bee diseases indiscriminately around the globe it would be hard to come up with a better system. Of course the companies involved do their best to combat disease. Major outbreaks in their factories would be enormously costly; all of the factories I have seen are clean, with protocols in place to reduce the spread of any diseases that appear. Sickly nests are destroyed, and samples of bees are regularly screened. Nonetheless, when scientists examine the nests that leave these factories, they often report them to be infected with a range of diseases. I am currently co-supervising Pete Graystock, a PhD student based at the University of Leeds in the lab of Bill Hughes (himself a former PhD student of mine from my Southampton days). Pete has been using genetic tools to detect the DNA of parasites in commercial bumblebees, which is the most sensitive technique currently available. He has examined nests from the three main companies that supply bumblebees to the UK, and has found *Nosema bombi* and *Apicystis bombi* to be common in nests from all three. These are nasty diseases, which can readily kill their hosts. He also found *Crithidia bombi* in some, and Deformed Wing Virus in others. If Pete's data are correct, this is pretty damning evidence that, despite the extensive precautions used in the factories, their bumblebees are riddled with diseases.

If these commercial bees are infected, then just as the bees themselves are certain to escape, so are their diseases. If these escape into a part of the world where they do not naturally occur, they could have a devastating impact on native bees that may lack resistance to them. One of the major reasons for the collapse in populations of Native Americans following the arrival of Europeans was their exposure to European diseases. Influenza, chicken pox, measles and the like, which rarely prove deadly to Europeans, caused devastating loss of life amongst Native American tribes. Interestingly, exactly the same thing might have happened with North American bumblebees. In the 1990s, queens of various North American bumblebees were taken to Europe and reared in factor -ies alongside the European buff-tails. The established nests were then returned to North America. Shortly afterwards, the western bumblebee, yellow-banded bumblebee and rusty-patched bumblebee, all widespread and common species, suddenly disappeared from much of their range. These species are all closely related. Their only other close relative in North America, Franklin's bumblebee, was always very rare but now seems to have disappeared entirely. Intensive searches for it in sites where it used to occur in northern California and Oregon have failed to find a single one since 2006, so it may have gone globally extinct. An entire group of closely related bumblebees has been devastated across a continent in the space of a few short years. To put this in context for European readers, this would be the equivalent of the disappearance of ubiquitous species such as buff-tailed and white-tailed bumblebees, the everyday bees that make up the majority of those on garden flowers. Many conservationists in North America blame these declines on the accidental introduction of a European bee disease along with the nests that were shipped in from Europe. *Nosema bombi* is a popular candidate. Or it may be a viral disease that we have yet to identify. In truth we may never know. Almost nothing is known about the bumblebee diseases that were present in North America before the 1990s, although once again this could be examined by looking for their DNA in museum bumblebees. Studying the few survivors is unlikely to be revealing as they are presumably those that didn't catch the disease, if indeed it was a disease that wiped them out.

Whether or not a European disease is the cause of these terrible declines, the principle remains. Shipping bees around is inherently risky unless they can be guaranteed to be free of disease. Oddly, despite the commercial trade in bumblebees now being well over twenty years old, there is very little regulation. Honeybees cannot be transported between most countries unless they have been certified free of an agreed list of their major diseases, but no such regulations have been applied to bumblebees. In the UK, there are no independent checks whatsoever on the bumblebees that are shipped in, despite Defra being well aware of the situation – they paid me to provide them with a report on the issue in 2009.

All of this is of relevance not just to bumblebee conservationists and those who supply and use bumblebees for pollination. There is a bigger issue here. Honeybees have also suffered from major health problems in recent years, both in North America and Europe. One of the most significant of these is caused by the parasitic mite, *Varroa*. First discovered in Asia, this unpleasant creature sucks the blood of honeybees and their brood, and in doing so rapidly spreads viral disease within colonies. Unlike their Asian cousins, European honeybees – which are also the bees kept in North America – have very little resistance to *Varroa*, and it can rapidly destroy their colonies. It was accidentally introduced from Asia to Eastern Europe in the 1960s, and has since spread steadily and relentlessly westwards, arriving in the UK in 1992. It also turned up in North America in 1987 and New Zealand in 2000. The only country that it has yet to conquer is Australia. Beekeepers have been battling with *Varroa* ever since.

In 2007, a new honeybee plague struck North America. During the winters of 2007 and 2008, commercial beekeepers in the USA lost between 30 and 90 per cent of their honeybee colonies. The symptoms were rather peculiar: there were no corpses. The adult bees had simply disappeared, leaving behind tens of thousands of empty honeybee hives. Various terms were coined for this phenomenon, my favourite being Marie Celeste Syndrome, but these days it is generally known by the clumsy name of Colony Collapse Disorder, or CCD for short. Beekeepers in Europe heard about the catastrophe in the USA, and when rumours of heavy colony losses in the UK surfaced in 2008 there was something approaching hysteria at the prospect that CCD had crossed the Atlantic. There was wild speculation as to the causes – disease, pesticides, intensive farming, GM crops, even mobile phones were variously blamed, and the media had a field day. In fact the rumours of colony losses in Europe were rather exaggerated; and often when colonies did die it was due to obvious causes and not CCD. It is unclear whether CCD affected honeybees in Europe at all, not least because we still have little idea what CCD actually is. In the USA scientists have been frantically searching for the cause, but five years later we are not much wiser, and the heavy colony losses seem to have ceased, or at least declined. Most experts think that there isn't a single cause – that it probably involves one or more diseases, perhaps viruses, but that some other factors such as exposure to pesticides may trigger outbreaks.

To provide a bit of perspective on CCD, it is worth noting that it is normal for perhaps 10 to 25 per cent of honeybee colonies to die every winter, due to a variety of causes. Also, CCD is probably not new. There are records from 1869 of outbreaks of 'disappearing disease', which certainly sounds very similar, making the mobile phone theory look rather flimsy.

Whatever the truth behind CCD, there is no denying that honeybee keepers the world over are having a tough time, and that diseases of one sort or another are a major part of their problems. How does this relate to the commercial trade in bumblebees? To rear bumblebee colonies, you need pollen – lots of it. To rear hundreds of thousands of bumblebee colonies, you need lorry-loads of pollen. One million nests – a conservative estimate of the European trade – probably requires in the region of 500 metric tonnes of pollen each year. Unfortunately, there is only one way to get hold of these sorts of quantities of pollen – from honeybees. Pollen can be collected from honeybee hives by fitting a metal grille to the entrance. The grille has holes that are just large enough for honeybees to squeeze through, but small enough so that the balls of pollen on the legs of returning foragers get knocked off into a collecting tray beneath. It seems a little cruel as the poor bees spend all day foraging only to have the fruits of their labour repeatedly snatched away just as they get home, but it doesn't do them any real harm. Of course you can't attach one of these grilles to a honeybee hive for too long or the colony would run out of pollen and the brood would begin to starve. To obtain the quantities of pollen required, the factories must buy it from beekeepers all over Europe. Are the honeybee hives from which the pollen is obtained all free of disease? This seems highly unlikely, perhaps impossible, since almost all honeybee hives have some viral and fungal diseases. Hence tonnes of pollen, almost inevitably contaminated with a range of bee diseases, are shipped into the factories, and there it is fed to bumblebees, which are then despatched all over the world. We know that bee viruses such as Deformed Wing Virus will readily infect both bumblebees and honeybees, so it is no surprise that Pete Graystock found it in commercial bumblebee nests. Honeybee diseases to which bumblebees are immune may also be spread, simply because the bumblebee nests are shipped out with a supply of pollen inside them, and as soon as the bumblebees are deployed at their destination the workers will fly out and start visiting flowers, perhaps carrying contaminated pollen on their bodies.

A lot of time and money is spent on trying to control and manage the spread of honeybee parasites and diseases; many countries including the UK employ bee inspectors to keep an eye on the health of honeybees in their area. At the same time, almost no attention is being paid to the mass transport of bumblebees. It is quite likely that the bumblebee trade has led to the wholesale redistribution of bee diseases around the globe, including those that infect honeybees. It may have had something to do with CCD in honeybees. Parasitic mites from Europe have been accidentally spread to Japan with commercial bumblebees, and they now attack native Japanese bumblebees. In Chile and Argentina there is a strong suspicion that non-native diseases that arrived with imported bumblebees are responsible for rapid declines of the giant native *Bombus dahlbomii*, the only bumblebee that is native to southern South America. *Crithidia bombi* and *Apicystis bombi* may not be native to South America, but they are rapidly spreading with introduced buff-tails from Europe and ruderal bumblebees from the UK, along with who knows what else. Buff-tails could easily spread through much of South America, using the cool temperatures found at altitude in the Andes to spread northwards towards the equator, and if other native South American species respond in the way that *Bombus dahlbomii* has, then it is conceivable that many of them could be wiped out.

I don't want to paint a picture of the commercial bumblebee breeders as irresponsible cowboys. Of course they are in the business of making money, but it is not in their interests to spread bee diseases, or to have non-native bumblebees escape into new environments. Most of these companies also supply biological control agents, and actively promote them as alternatives to chemical pesticides. Many of the staff involved in bee-rearing are passionate about bumblebees, and mortified at the suggestion that they might be doing harm. Two of the bigger companies recently started rearing native buff-tails (*Bombus terrestris audax*) for the UK market, following criticism that they were shipping in non-native bees. Nonetheless, there is mounting evidence that current practices are threatening the health of wild bees the world over. It is high time that strict hygiene regulations were imposed on the bumblebee trade, before any more disasters occur. It would be better still if local, native bees were reared in factories in the country where they are to be used, negating the need for long-distance transport. The native buff-tails being reared for use in the UK are currently bred in factories in mainland Europe and shipped in. Some countries, such as Canada, New Zealand and Turkey, have banned importation of bumblebees. This has forced companies to set up local factories producing local bees, something they are not keen to do since it is cheaper to have just one large factory and distribute bee nests from there. Unfortunately, the UK government seems loath to go down this route, despite the environmental benefits and the opportunity to create jobs.

Is there an alternative to using commercial bumblebees to pollinate crops? For tomatoes in glasshouses, the answer is probably no. Farmers would certainly not want to return to using teams of labourers with vibrating wands. However, commercial bumblebees are increasingly being used to pollinate outdoor crops such as strawberries, blueberries and apples. Farmers used to rely upon wild bees to visit such crops, but are increasingly of the opinion that there are not enough wild bees to go around. In the pear orchards of Sechuan in China, intensive farming has all but eradicated wild bees, and the farmers now pay locals to clamber amongst the trees each spring armed with a paintbrush and a jar of pollen, with which they hand-pollinate every flower. This is just about viable in China, where labour is plentiful and very cheap, but it is not an option for most farmers elsewhere. In Canada, intensive blueberry farming over vast areas has also led to low populations of native bumblebees, for there are few wild flowers for them to feed on when the blueberries themselves are not in flower, and there are also few places for them to nest. Just as in China, widespread pesticide use no doubt exacerbates the problem. Many blueberry farmers now buy in commercial nests of the native common eastern bumblebee, *Bombus impatiens*, to pollinate their crop. On the soft-fruit farms of Perthshire, most farmers now buy in dozens, in some cases hundreds, of buff-tail nests each year. The majority are placed in polytunnels to pollinate raspberries, while some are used outdoors for strawberries. The raspberry polytunnels have open ends and the plastic on the sides is rolled up in summer, so both raspberries and strawberries could be pollinated by wild bumblebees if there were enough. Clearly most farmers think that there are not.

I was a little suspicious about this. On the one hand, it is hard to imagine a Scottish farmer spending £40 per nest on bumblebees if he didn't have to. On the other, strawberry and raspberry yields vary a lot from year to year, and it would be very difficult for a farmer to know for sure what benefit he was getting from commercial bumblebee nests unless he did a proper experiment. There are umpteen examples of clever marketing persuading folk to buy products that are no use whatsoever. The cosmetics industry depends upon it.

So it was that in 2010 I employed one of my former PhD students, Gillian Lye, to look into this, funded by a little money I had left over from a previous grant. She placed four commercial bumblebee nests next to a half-hectare plot of raspberries near Dundee, mimicking the recommended density of six to nine nests per hectare of crop. She opened and shut the doors on the nests at weekly intervals so that we could compare fruit set and raspberry yield in weeks with and without the aid of commercial bees. Lots of wild bumblebees came to pollinate the raspberries – white-tails, buff-tails, early bumblebees, common carders, even the beautiful bilberry bumblebees which have enormous red bottoms. (I admit, an enormous red bottom doesn't sound beautiful, but reserve judgement until you've seen one.) Quite a few honeybees turned up, presumably from nearby hives. Yet despite all these bees, the yield increased on average by 8.3 per cent when the commercial nest boxes were opened. I must confess that I was dismayed by this result, although I shouldn't have been. Scientists are supposed to do experiments without looking for a particular result; we are meant to strive to be impartial at all times, else we risk subconsciously biasing our results in some way. Nonetheless, I had hoped that the commercial bees would prove to be unnecessary; if they had, I could have used the evidence to persuade farmers not to buy them.

Of course 8.3 per cent doesn't sound very much, but when translated into cash the figures are more impressive, for raspberries are a valuable crop. Gillian estimated that the commercial bumblebees increased yield from this half-hectare plot by 63 kilograms per week, and since the colonies last about six weeks this equates to an extra 378 kilograms of raspberries, worth approximately £2,259, for an outlay of about £160. Of course commercial farms are much larger than this, and the larger the stand of crop, the less able wild bees are likely to be to provide anywhere near adequate pollination, so on the basis of this study, purchasing commercial bumblebees looks like a very sensible option for raspberry farmers.

On the other hand, this is not the only option. Gillian's data provides clear evidence that, just as in Sechuan and parts of Canada, we no longer have enough wild bees to pollinate our crops, but that probably wasn't always the case. Not long ago we had many more bees, and farmers managed very well without buying in extra. We used to have more bees because farming was different – farms had more flowers, and used fewer pesticides. Commercial bumblebees are expensive, so that large soft-fruit farms are spending several thousand pounds per year on them. What if this money was spent instead on boosting wild bumblebee numbers, by planting strips of wild flowers, or providing them with nesting habitat? Could this provide a more effective and infinitely more environmentally friendly alternative to using commercial bumblebees, at least for outdoor crops?

I don't yet know the answer to this question, but two of my PhD students are trying to find out. Ciaran Ellis and Hannah Feltham are working together to evaluate the effectiveness and economic outcome of alternative strategies. Planting strips of wild flowers near fruit crops ought to increase wild bumblebee numbers, but it might draw bees away from the crops, which is the last thing the farmer would want. It also requires money for seed, diesel for the tractor to prepare the ground, sow the seed and manage the strips, and farm labour. On the other hand, if it works and is cost effective it would be marvellous to be able to go to farmers with evidence that they can make more money by growing wild flowers on their farm.

I think it is unlikely that we can come up with a scheme that would completely do away with the need for commercial bumblebees for raspberries in polytunnels, not least because some varieties are grown that flower in March and April, when very few wild bumblebees are on the wing in Perthshire. But we might be able to reduce the number of nests that have to be bought, and perhaps do away with the need for them entirely in July and August when wild bumblebees are most abundant.

Aside from the costs and risks associated with commercial bumblebees, there is one more good reason for farmers to ensure that they do not forget about the services offered for free by wild bees. Most commercial bumblebees come from a very small number of factories, and only one species is available in Europe. If anything should happen to the supply of commercial buff-tails, such as a major outbreak of a disease in one or more of the factories, then many farmers would not be able to get hold of them and the price per nest would skyrocket. If there were also no wild bees, then crops would fail, and some farmers would go bankrupt. This is not just idle conjecture – Colony Collapse Disorder in the USA led to a massive rise in the cost of hiring honeybee hives for crop pollination because so few were available, and hence had a huge financial impact on farmers. Depending entirely on one commercial species is an inherently risky strategy; putting all one's eggs in one basket. Wild bees can be viewed as a backup strategy, an insurance policy in case supplies of commercial bees should fail.

When you next squirt Heinz tomato ketchup on to your fish and chips, reflect on the nature of the modern world. Your ketchup was most likely made in a factory in the Netherlands from tomatoes grown in Spain, pollinated by Turkish bees reared in a factory in Slovakia. I'm sure that our food supply chain doesn't need to be quite so convoluted. You might also reflect that every cucumber, aubergine, runner bean, blackcurrant and pepper that you eat was almost certainly pollinated by a bumblebee, perhaps reared in a factory, or perhaps a wild bee. A tin of baked beans largely comprises navy beans that were pollinated by bumblebees, and a sauce made from bumblebee-pollinated tomatoes. We owe these little creatures for all that they give us . . .

CHAPTER FIFTEEN
Chez Les Bourdons

To make a prairie it takes a clover and one bee,
One clover, and a bee,
And revery.
The revery alone will do,
If bees are few.

Emily Dickinson

There are two spiritual dangers in not owning a farm. One is the danger of supposing that breakfast comes from the grocery, and the other that heat comes from the furnace.

Aldo Leopold
(American environmentalist)

I have always hankered after owning some land that I could manage as my own private nature reserve. Gardening for wildlife has been a lifelong occupation, but I really fancied something bigger and more ambitious than my quarter-acre suburban garden. On an academic salary, buying a substantial area of land in the UK was not an option, so I found myself looking further afield. At the time, in 2002, I lived in Southampton, and with a ferry port nearby in Portsmouth the obvious option was France.

I spent many happy hours surfing the Internet, looking at all manner of fantastic French properties. I failed to resist clicking on the link for the dilapidated twenty-bedroom chateau near the Loire, available for the price of a three-bed semi in the home counties, or looking longingly at the pictures of a rustic farm with 400 hectares of mountainous forest and scrub in the Cévennes, but neither was terribly practical and, relatively cheap though they were, they were beyond my very limited budget. In the end I narrowed my search down to the Charente, a peaceful backwater of rural France between Limoges and Poitiers, about halfway down France and west of centre. I'd holidayed in the area as a boy – my family invariably spent two weeks camping in France every summer – and I remembered catching exquisitely beautiful white admiral butterflies in the lovely oak forests with which the area abounds, as well as spending many happy days rock-pooling and playing French cricket on the great beaches of the west coast. Property also happened to be absurdly cheap, and the area is within a day's drive of the Channel ferry ports.

So it was that in a cold and wet October I found myself on a one-week whistle-stop tour of estate agents in the Charente. I'd booked in advance with them, and arranged to view a number of properties that I had seen on the Net and which looked of interest. My father came with me – I think my wife Lara had had a word with him and given him strict instructions to prevent me from buying anything too ridiculous. It was a hectic schedule – we visited a different agent in a different town every day, and then spent our evenings driving on to the next town. Most of the properties I had seen on the Internet had been sold years earlier – it seemed that French estate agents updated their sites only every decade or so, such is the pace of life in rural France.

The French countryside is absolutely littered with beautiful but dilapidated and neglected old houses, usually built of stone, with heavy oak beams and hand-made clay tiles. Many are already more or less beyond repair. In England these old piles would fetch an arm and a leg and perhaps a kidney too with folk paying a premium for the privilege of spending a year or two in a mobile home in the garden while having them reconstructed – but in France they cost very little and even at such low prices they don't sell. There is not much rural employment, and in any case it seems that the French don't want these draughty old properties, understandably preferring to live in cosy modern housing. Most sales of older, rural properties are to Brits looking for an idyllic holiday home or a new life in the country, but many of them end up back on the market within a few years as the upkeep is considerable, while those looking for a new life are often driven home by the difficulties of integrating into the local community, or the lack of viable ways to earn an adequate living. So it is that the market for rural property stagnates, and one by one the lovely old houses are falling down.

All of this was good news for me. I didn't plan to make a living in France – I just wanted somewhere to grow flowers and feed bees.

Despite my insistence that I had a very limited budget and was only interested in properties with land, the agents enthusiastically showed my father and me around all sorts of unsuitable properties, many on the verge of collapse, and often with only a small garden. Blind optimism and selective deafness seem to be prerequisites for employment as a French estate agent, perhaps because they sell so few properties. The many derelict and boarded-up heaps included an abandoned shoe factory somewhere near the town of Piégut, with machinery still in place, and a fourteen-bedroom chateau on the edge of Fontenay-le-Comte. The latter was being sold by an ancient couple who lived in just two rooms, all that they could afford to heat, while the rest of the building fell down around them. I felt very sorry for them, and it would have made a wonderful renovation project for someone with a bottomless pocket, but it was not what I was after.

On the fifth morning, after a night at a charmless motel called L'Escargot, built on the banks of a busy dual carriageway, we visited an agent in the oddly named village Champagne-Mouton. It was a damp, misty morning, and the estate agent's was closed when we arrived at the appointed time, so we went for a walk through the eerily silent back lanes of the village, startling a cow which was the only sign of life. When we got back the agency was open and we were briefly greeted by a rather unfriendly Englishman, but no sooner had we walked in the door than he muttered something about an errand and disappeared. His surly Russian wife spoke no English or French, so far as I could tell, but she made us a lukewarm instant coffee and then disappeared into a back room.

When the agent returned an hour or so later he seemed distracted and more than a little put out at the bother of showing us anything at all. In an unusual approach to his trade, and at marked odds with the approach of his French counterparts in other agencies, he tried to deter us from viewing any of his properties, emphasising their many failings, but in the end he begrudgingly agreed to show us a couple. The first was an old farm with 13 hectares near the village of Épenède, one of very few properties I'd seen on the Internet which was actually still for sale. The mist had not cleared by the time we arrived and, as we had come to expect, the place was in a parlous state. As we approached the end of the no-through lane that accessed it, we were greeted by a ferocious black dog that howled and hurled itself towards us, fortunately restrained by a long length of chain attaching it to a stone barn. Pieces of rusting, broken-down farm machinery littered the yard, many of them having lain there so long that they were encrusted with ivy, and had begun to sink slowly into the mud. The owner, a Monsieur Poupard, proudly showed us around.

The main house was a long, low, stone rectangle, with living accommodation on the ground floor and hay storage above. He lived in three small rooms, with only a stone sink to wash in – his loo was in an outside shed and consisted of a plank with a hole in it over a bucket. Poupard himself was a tiny, weather-beaten man in the ubiquitous blue overalls worn by all French farmers, with a flat cap pulled low over his eyes. He'd lived in the property all his life, but had clearly not maintained it. The window glass was mostly broken, with old fertiliser sacks nailed over the broken panes to keep the wind out. Successive generations of sacks had frayed and split in the wind, so he had nailed layer after layer on top over the decades. The front door was rotted away at the bottom, the gaps patched with flattened, rusty tin cans. The clay-tiled roof leaked so badly that there were pools of water on the floor – his old iron bed-frame was standing in an inch of fetid water. The house was surrounded by a collection of barns and outbuildings, most of which appeared to be in an even worse state of repair than the main house. The largest barn had holes in the tile roof through which an albatross could comfortably soar, and was clearly in imminent danger of collapse.

As we explored the upstairs of the main house, with Poupard taking care to steer us around the more rotten floorboards for fear we might fall straight through into the pool in his bedroom, we startled a barn owl from its roost on an old beam. I excitedly ran down the stairs and outside to watch as it circled the house; it eventually alighted in an oak tree next to the drive. I hadn't seen a barn owl in years. Poupard followed me out, looking at me rather curiously. He had obviously not expected me to be quite so interested in owls. Since we were outside, he pointed out the boundaries to the property – distant lines of oaks barely visible through the mist. The farm was situated in the midst of a sloping field which to me, someone who had never previously owned more than a quarter of an acre, seemed unimaginably vast.

I can't entirely explain why, but I was enchanted. I think the owl swung it. I guess the asking price had something to do with it, since it was of such a paltry magnitude that it wouldn't have been enough to buy a studio flat in Falkirk. My father did his level best to talk me out of it, and clearly thought that I had lost my marbles. He kept rubbing his head, which is a sure sign that he is worried. Undeterred, after sleeping on the decision, I offered the asking price the next day and Poupard accepted with alacrity. The legal formalities took a few months, and in February 2003 I drove back to sign the paperwork. After the signing I went to look over my new property, and I must admit that I began to have second thoughts. In the depths of winter it was colder, damper, and more formidably uninhabitable than ever it had appeared before. Icy rain was sheeting down from the west. Lara was with me, six months pregnant with our second boy Jedd, and she narrowly missed being savaged by the baying hound that threw itself towards her until violently pulled short by its chain. Finn, our eldest, was just a toddler and he backed away from the dog, tripped over on some rusting metal and fell in the mud, where he proceeded to burst into tears. Lara had not seen the property before and was, understandably, less than impressed.

I returned in March with my father for a ten-day blitz to make the place a little more habitable, and I have been slowly working on it whenever I get the chance ever since. My father often comes down with me to help out and, although his DIY skills are woeful, it is great to have the company. We now have, wonder of wonders, a flushing indoor toilet, sinks with hot running water, even a shower, and a roof that leaks only occasionally. Poupard would scarcely recognise it.

On those first trips down with my father we camped in the garden – much more pleasant than attempting to sleep in the house. For a toilet, we simply took a spade into the edge of the field and dug a hole – far more appealing than Poupard's external facilities. On one such occasion, late in the first summer after buying the property, I was squatting behind the hedge, surveying the rather splendid view across the fields, when I heard a distinctive high-pitched buzz. There were tall pink willowherbs in flower along the field margin, and there, foraging, I was delighted to see a shrill carder bumblebee. It was a rather beautiful male, with straw-yellow stripes and a reddish bottom, busying himself collecting nectar. This species is a great rarity in the UK – I had only ever seen them on the wilds of Salisbury Plain and in the Somerset Levels, and even there they are hard to find. To have them living on my farm in France was wonderful, and as soon as was convenient I rushed to tell my father.

Chez Nauche – the official name of the farm – is the most peaceful place. Outside the front door, the only evidence that there are other humans in the world are the vapour trails in the blue sky above, and the occasional distant chug of a farmer's tractor. Wall lizards scurry along the south-facing front of the house, snatching flies and fighting territorial battles. At night, garden dormice – beautiful and incredibly agile creatures with the face markings of a miniature badger – churr at one another while eating grapes from the vines I have trained over the walls, while glow-worms emit their phosphorescent green light from among the cracks in the old flagstone path. My boys have grown up spending a month-long holiday here every summer, catching mantises and butterflies in the meadow, building tree houses, and swimming in the local lakes and rivers when the afternoon heat becomes too much.

The changes I have wrought on the house are nothing compared to what has happened with the meadow. When I bought the property, the land had been under arable crops until very recently. Poupard had put the field down to grass the year before, but it was more or less a monoculture of false oat grass, a tall, coarse, emerald-green grass. Old plastic sacks of ammonium nitrate were strewn around in the barns, so it was clear that the fields had been fertilised, although I doubt Poupard would have put on more than the bare minimum as fertilisers are costly. Nonetheless, even small amounts are disastrous for flowers – they encourage coarse grasses to grow fast and tall, squeezing out all but the toughest herbaceous plants. Only along the edges of the drive and the hedge bottoms were there any flowers, and hence there were few bees and butterflies.

There is no easy way to reduce the fertility of soil in a meadow, yet it is impossible to recreate a rich flower community without first doing so. One option is to scrape off and remove the topsoil, but for such a large area this would have meant removing thousands of tonnes of soil and hence would have been prohibitively costly, as well as posing the considerable problem of where to dispose of it. An alternative is to deep plough, a tricky technique whereby the ploughman flips the top 2 feet or so of soil upside down, burying the more fertile topsoil under a thick layer of less fertile subsoil. Both methods require that one then sow a wild-flower mix, preferably using locally sourced seed. Such seed mixes cost several thousand pounds per acre; enough to sow the fields at Chez Nauche would have cost considerably more than the property itself.

There is a third option, but it requires considerable patience. If the hay is cut and removed each year, then some of the fertility is removed with it. So long as no more fertilisers are added, year-on-year removal of the hay results in a slow decline in fertility, and allows wild flowers to gradually creep back. I managed to contact a local farmer, Monsieur Fonteneau, who maintains a herd of 350 hungry goats, and he agreed to harvest my hay each year to provide winter fodder for his animals. Every July, he and his sons cut the hay and bind it into vast cylinders, removing about

80 tonnes of material and with it a small amount of the nitrogen and other artificial nutrients added by Poupard. Each year, out of politeness, Monsieur Fonteneau will pop in to exchange pleasantries. He is usually accompanied by one or more of his rotund boys whom he is training up to take over the family business. I offer them each a bottle of cold beer, which they drink while I rapidly exhaust my feeble repertoire of French farming-related conversation. None of them speaks a word of English, or if they do they have never let on, and my French is dimly remembered from lessons at school. To prepare for Fonteneau's annual visit I have rehearsed a number of conversational gambits, such as '*Le foin est bon cette année?*' (The hay is good this year?) and '*Comment sont les chèvres?*' (How are the goats?). Of course I rarely understand the answers, although I believe that his goats may have recently suffered from some sort of bloat. It is all a little awkward, but fortunately Fonteneau *et fils* drink fast; we smile, we shake hands, and we all breathe a sigh of relief that the conversation is over for another year.

Very slowly, over the last decade, the meadows have begun to fill with flowers. It has been frustratingly slow. Even if conditions are suitable, the flowers have to come from somewhere, and this limits how fast they return. Some, such as poppies, can survive for years as seed in the soil, but most of the perennial plants that typically thrive in a flower-rich meadow cannot, so seeds have to arrive from elsewhere. Some flowers have seeds which readily disperse over a huge distance. Dandelions are a well-known example – their seeds hang below a fluffy pappus of fine hairs and can drift for miles on the wind. They have lots of relatives – mouse-ears, cat's-ears, hawkbits and so on (the last taking its name from an ancient belief that hawks ate them to improve their eyesight). All share the same seed-dispersal mechanism, and so these were among the first flowers to arrive. The meadow is now a sheet of yellow in June and July, but only from late morning through to the afternoon, for these flowers close overnight and they are not early risers. Wild carrot followed close behind; the seeds are light and flattened, and so can blow a short distance on the wind. They seem to be thriving in the drier soils in the higher parts of the meadow. Other plants are much slower. Cowslip seeds are heavy and round, so they will always fall close to their parent. There were some cowslips along the side of the track to the farm, and these have been slowly spreading into the meadow, but they have only made it a few yards so far.

From a bumblebee's perspective, legumes are among the most vital components of a wild-flower meadow. Plants of this family include clovers, trefoils and vetches, as well as garden vegetables such as peas and beans, and they have an unusual trick that allows them to thrive in low-fertility soils. Their roots have nodules, small lumps inside which live *Rhizobium*, bacteria that can trap nitrogen from the air and turn it into a form usable by plants. Most plants are severely limited in their growth by a shortage of nitrates, which they require to construct proteins, and yet nitrogen, the element from which nitrates are made, comprises 80 per cent of the air that surrounds them. By enlisting the help of bacteria, legumes get around this problem – they feed their pet bacteria on sugars, which they obtain by photosynthesising, and in return the bacteria provide them with nitrates. This relationship gave legumes a huge advantage in the days before artificial fertilisers were widely deployed. Ancient hay meadows are full of clovers, trefoils, vetches, meddicks and melilots, able to outcompete grasses because they alone have access to plentiful nutrients. Most of these plants are pollinated by bumblebees.

Over the last fifteen years or so I have gathered a lot of records as to which bumblebees feed on which flowers, separated into visits for pollen and visits for nectar. One of the striking features of these data is that some bumblebee species, such as the brown-banded carder, the garden bumblebee and the ruderal bumblebee, seem to get nearly all of their pollen from legumes. What is so special about legume pollen? To find out, I gathered pollen samples from various different flowers – a remarkably tricky business, which bees make look ridiculously easy – and sent them off to a lab in Cambridge where their nutritional composition was analysed. It turned out that legume pollen is especially rich in protein. What is more, the protein in legume pollen was itself unusually rich in 'essential amino acids', those which animals cannot synthesise for themselves. To ensure the fidelity of bumblebees, it seems that legumes offer pollen which is unusually rich in high-quality protein. Since pollen is the only source of protein available to bees, it makes perfect sense for them to selectively visit the flowers that provide the richest source.

There is an interesting parallel to be made here with vegans and vegetarians. Bees are the vegetarian descendants of wasps, having turned to feeding on pollen and nectar instead of the flesh of animals. They visit legumes to obtain protein. In much the same way, human vegetarians and especially vegans tend to include a large proportion of pulses in their diet because they provide a rich source of protein and particularly of essential amino acids – and pulses are the seeds of legumes. Legumes can afford to put plentiful protein into both their pollen and their seeds because of their root nodules.

The mutualistic relationships between bees, the flowers that they pollinate, and the bacteria that live within the roots of those plants are at the heart of the functioning of a natural, species-rich meadow. The problem is that these relationships can be ruined by application of a sack of fertiliser, which allows the grasses to swamp the legumes and other wild flowers, swiftly resulting in a bright green, flowerless sward, with no legumes, no *Rhizobium*, and no bees. In the farming world this is known as 'improved' grassland. In the 1940s, Britain had in the region of 15 million acres of flower-rich grasslands. It is hard to get precise figures, but about 250,000 acres remain; a staggering loss of over 98 per cent. Fertilisers were cheap, and successive governments were keen to persuade farmers to boost productivity, so ecosystems that had taken hundreds of years to develop were subject to swift and wholesale destruction. Most of what remains of our flower-rich grasslands is in tiny patches of less than 5 acres, usually on nature reserves. I cannot find equivalent figures for France, but I suspect that they would be equally depressing for lowland areas such as the Charente as I have yet to discover a single sizeable patch of flower-rich grassland anywhere near Chez Nauche other than my own. It is hardly surprising, then, that many bumblebees and other insects have disappeared from much of the countryside, both in Britain and across Western Europe, for most of the flowers, including their favourite foods, have all but disappeared.

Unfortunately most legumes, like cowslips, have heavy seeds and no clever mechanism for dispersing them on the wind, and so they are slow to return once they have been lost. Many, such as vetches, have chunky pea-like seeds, so unless they are carried by ants – which they sometimes are – they do not disperse more than a foot or two each year. To speed them up a bit, I have cheated a little in my meadow by collecting red clover and bird's-foot trefoil seed from nearby verges and sprinkling it by hand. Whether this has helped is hard to say, but both species are now scattered in patches through the sward, and each year the patches spread further. Clumps of bush vetch have appeared, as have other legumes such as hare's-foot clover. Slowly, the false oat grass has declined, presumably because the nutrients it needs are disappearing.

Aside from creating my own bumblebee reserve, I had another motivation for wanting to buy some land. Much ecological research requires long-term experiments, but long-term experiments are very hard to maintain. Grants for research generally run for only three years, and cash-strapped universities rarely have sufficient land that they can set aside over many years for such purposes. I had my fingers burned in this respect at the University of Southampton. I set up an experiment to study hybridisation between red and white campions in a walled garden owned by the university.* It involved planting out dozens of patches of plants, either red, or white, or mixtures of the two, and then studying them over time to see whether red or white or hybrid offspring came to predominate. These are fairly long-lived plants so I planned to run the experiment for at least ten years – only to have the experimental plots sold off to developers after just eight, before the most interesting results could be obtained. If I owned the land myself, such disasters could be avoided.

Recent research by Richard Pywell and colleagues at the Centre for Ecology and Hydrology in Oxfordshire suggests that the restoration of species-rich meadows can be accelerated by use of yellow rattle, a hemiparasitic plant. Yellow rattle is a relative of foxgloves, but is a plant typical of ancient hay meadows. Much-loved as a nectar source by long-tongued bumblebee species, it is a pretty but unpretentious annual that grows to about a foot or so tall,

* Campions are fascinating for a range of reasons. Each plant is either male or female, unlike most plants which are hermaphrodites. They suffer from sexually transmitted diseases, fungi with the oddly appropriate name of smuts, the purple spores of which are spread from flower to flower by bees, and which force infected female plants into a transsexual imitation of a male. Red and white campions will hybridise to produce pink offspring, but some unknown mechanism manages to prevent the two species from merging into one – the subject of my long-term and prematurely aborted experiment.

with small yellow flowers. Its name derives from the seeds, which rattle within the dried pods. Yellow rattle is unusual in that it parasitises other plants, particularly grasses, tapping into their root systems and extracting nutrients.* Since the grasses are being parasitised, they grow less, leaving more room for other flowering plants. Pywell demonstrated that sowing rattle seed into an English meadow significantly boosted the diversity of flowers present by suppressing growth of grasses.

This seemed like an interesting idea to follow up in France, since adding rattle might both directly provide flowers for bees and boost the colonisation process. It occurred to me that there were many other hemiparasitic relatives of rattle that might be deployed instead of or as well as rattle, and that it would be worth trying out as many as possible to evaluate which were most effective. To this end, in September 2010 I arrived at Chez Nauche with a posse of helpers, volunteers from among my PhD students and the staff of the Bumblebee Conservation Trust. We marked out 120 plots, each 10 by 10 metres square, using aluminium plates pegged to the ground. Before sowing the seeds we had to prepare the ground, which first involved cutting and removing as much vegetation as possible. I have an ancient ride-on mower which I used at first, but after half a dozen plots it spluttered to a halt, and for the life of me I couldn't persuade it to go again. I rushed

* Some plants, such as the bird's-nest orchid, are entirely parasitic, gaining all their nutrients from other plants. They have no need for chlorophyll since they do not bother to photosynthesise themselves, and so are usually an anaemic pale brown in colour. Yellow rattle and its relatives are described as hemiparasitic because they have a mix-and-match strategy; they do make their own energy via photosynthesis and so they need green leaves, but they also steal nutrients from other plants when they can.

out and bought a push mower, the largest I could afford but ridiculously small for the task in hand, and I spent the week mowing. Behind me, the rest of the team followed, raking, sprinkling seeds, and stamping them in with their feet. We used mixtures of seeds including two species of yellow rattle, two of cow-wheat, eyebright, and red bartsia, all of them hemiparasites, and we included control plots in which nothing was sown.

A farm track runs along the north boundary of the field, and on the second day I noticed Monsieur Fonteneau driving slowly past with one of his large sons. Both of them were staring at us intently, presumably wondering why this eccentric Englishman was mowing neat squares into the huge field with a small garden lawnmower, while a gang of helpers raked, sprinkled and performed strange shuffling dances. Over the following days other local farmers came to watch in bemusement. None of them came over to ask what we were doing, which was a relief as my French would not have been up to the task. I feel sure that our behaviour was the subject of considerable speculation in the nearby village bars.

My long-term aim is to follow these plots for many years, assessing the diversity of flowers in each. In spring 2011, eight months after we had set the experiment up, I returned to see whether the seed had taken. I could hardly bear to look as I drove down the track to the farm, but of course I did. At first glance there wasn't much to see. Over the next week, and with help from my PhD student, Leanne Casey, and Andreia Penado, a visiting Portuguese scholar, we identified and recorded the amount of every plant species in every plot, eagerly looking for the hemiparasitic species. We did the recording blind, in the sense that we deliberately did not check which mix had been sown in which plot before we evaluated them – otherwise we might have subconsciously biased our efforts, for example by looking harder for particular flowers in the plots in which they had been sown. One of the advantages of having a rather poor memory is that there was no chance that I would have remembered which were which from the previous year.

I was thrilled to find rattle in some of the plots, although there wasn't very much, but try as we might, we could not find any of the other hemiparasites. When I subsequently checked, the rattle had at least appeared in the plots in which it had been sown. No other differences were evident between the plots, but of course we wouldn't have expected them so early. It will take a year or two for the hemiparasites to become established in number, if they survive at all. I returned in 2012 and the pattern was much the same. In another four years or so I will be able to say which of the hemiparasitic species is best – at the moment, my money would be on rattle, which would be a slightly boring result since it will not add much to what Pywell and colleagues have already demonstrated. Nonetheless it will be valuable to show that the other hemiparasites don't work, if this is the case, so that other folk don't bother to try the same thing.

It has been incredibly rewarding to watch the large but rather dull grassy field that I bought a decade ago slowly turn into a flower-rich sward. Every year I discover that a few more flower species have somehow arrived. I keep a running total, which is now in excess of 130 plant species if I include the shrubs and trees in the field margin, and I am sure that there are more to find.

With the arrival of plentiful flowers, a myriad of insects have returned. Most insects can fly, and so it does not take them long to find good habitat when it is present. The field is now alive with butterflies every summer, including many which I would have regarded as spectacular prizes in my butterfly collecting days. Swallowtails soar powerfully over the field, searching for fennel on which to lay their eggs. Scarce swallowtails, with more angular wings and a slightly scorched appearance that gives them their

French name *le flambé*, swoop around my peach trees. Their larvae eat the leaves, which doesn't do the trees any good, but I can't begrudge such beautiful creatures. Glanville fritillaries lay batches of eggs on the plantains in the meadow. Their jet black larvae are gregarious, spinning dense webs in which they live together through the winter. On one memorable occasion a male purple emperor soared past, his iridescent wings flashing in the sunlight.* So far I have recorded over 60 species of butterfly, almost as many as live in the whole of the British Isles.

Among the more spectacular insects, praying mantis lurk on the tall stems of the wild carrots, attempting to snatch passing flies. Male great green bush crickets sing incessantly and deafeningly through the warm days of summer, in an attempt to impress a mate. On the ground, chubby black field crickets sit outside their burrows, their more melodic chirrups drowned out by the bush crickets. They have enormous oversized heads, which make them resemble cartoon characters. On a warm summer's evening, male stag beetles fly clumsily past, their wings barely able to carry the weight of their heavily armoured bodies. They seem to suffer an appalling mortality rate, for the local kestrels snatch them from the air and settle on the hay bales in the field to dismember them. In some years every bale has two or three stag beetle heads on it – the heads seem to be too tough to eat, so are discarded. Distressingly, the heads remain alive for a day or two, their antennae twitching and their great jaws slowly opening and closing.

* These spectacular insects normally hang around the tops of oak trees, and so are seldom seen. One old-fashioned technique that was used by butterfly collectors was to place a well-rotted dead rat on a woodland ride. Beautiful though the butterflies are, they have a macabre taste for the juices that leak from such a corpse and are often lured down.

Kestrels are not the only predators to take advantage of the bonanza of prey. Voles and mice have proliferated and attract hen harriers, which hunt low over the meadow. The local barn owl population appears to have boomed judging from the rate at which their pellets are deposited in my attic and the surrounding barns, although I rarely spot more than one or two of them. Impressively long western whip snakes bask on the stone walls of a collapsed barn to warm themselves in the morning, before stalking through the long grass for small mammals. They are a lovely dark olive with bright green speckles, almost iridescent in the sunlight. They live up to their name and can move like lightning so, as my boys and I have discovered, they are a great challenge to catch.

Even these fierce predators appear to have an enemy, a top predator that stalks the farm at night. I have found the remains of kestrels, snatched from their roost under the eaves of the house, and on one occasion I found the wing of a barn owl beneath one of their favourite perches – the rest of the bird having presumably been consumed. Most gruesomely, one early morning my boys discovered the remains of a particularly large and splendid whip snake which we had seen several times in the preceding days. All that was left of the 5-foot snake was a 6-inch portion from the middle – the remaining 4.5 feet having been eaten. I'm still not certain what this mysterious beast is. It can clearly climb, is nocturnal, and is big enough to eat a large amount of snake in one sitting. I have found scats that may belong to it, in appearance similar to those produced by pine martens with which I am familiar in Scotland. I've installed Velux windows above the bedrooms upstairs, and one morning awoke to find sizeable footprints across the glass – the beast had paced right above my head in the night. The prints were not crystal clear, but they roughly fitted with those of a large marten of some sort. Beech martens are said to be found in this part of France, so this is my best guess as to what it is, and I would dearly love to see it. Perhaps one day I will borrow one of Steph's bumblebee nest cameras and try to film its nocturnal activities.

Of course with all the flowers in the meadow, the bumblebees have also returned en masse. Shrill carders are now an everyday sight, hardly worth interrupting one's ablutions for. Red-shanked carders are also two-a-centime, distinguished from the even more common red-tailed bumblebees by the red hairy bristles on their legs. The long-tongued ruderal bumblebees remain scarce but regularly put in an appearance, feeding mainly on the red clover. All of these species are exceedingly rare in Britain, so it is a joy to see them thriving here. I have planted a row of lavender along the front of the house under the windows, so I do not even need to leave my picnic bench by the front door to see a plethora of butterflies, hummingbird hawkmoths and bumblebees. In July of 2009 I was enjoying a morning coffee when I spotted a slightly unusual-looking bee on the lavender. I think my heart knew what it was before my head had really had time to take in the details. It was, unmistakably, a short-haired bumblebee, the species we were working so hard to bring back from New Zealand to the UK. There are very few places in Europe where these bees are regularly seen, and to my knowledge western France was not supposed to be one of them but nonetheless here it was. Imagine my excitement, and the frenzied rush to grab my camera so that I could record the moment. I must confess that I have not seen one since then, but I hope that one day they might become regular residents.

The meadow still has a long way to go. It might take another twenty or thirty years, perhaps more, before it stops improving. I will keep adding to my plant, butterfly and bumblebee lists, and perhaps my boys will carry on when I'm not here to do it. To my mind, it is enormously reassuring to see nature reasserting itself.

Similar changes would occur in almost any meadow if it were left in peace for long enough, if it were not continuously ploughed and drenched with pesticides and fertilisers. Wildlife is resilient, it just needs a little space and time. Of course we can't turn the whole of Europe into a wild-flower meadow – lovely though that might be – but surely we can find room for a few more meadows like this one, where the multitude of interactions between bees, flowers, bacteria, snakes, mammals, birds and much else besides can be restored in all their glorious complexity?

I have a confession to make. I may argue that biodiversity is vitally important to our survival, that we need bees to pollinate our crops and so on, and that therefore we should set aside areas for conservation. But in reality, my motivation for restoring this particular meadow is much more selfish. It is, quite simply, the most beautiful place to sit on a summer's day. If you've never done it before, find yourself a wild-flower meadow – although they are scarce, there is probably one not too far away – go on your own, sit down in the long grass, and soak it all up. Take in the sounds: the buzzing of the bees, the chirping of the grasshoppers, the high-pitched screeches of fighting shrews deep in the grass, the melodic warble of skylarks silhouetted high above. Savour the scents: of basil and thyme and lady's bedstraw crushed underfoot, and of the flowers all around. Take in the gentle riot of colours: mauves, yellows, purples and pinks. It is as near to heaven as most of us are ever likely to get; what more reason could we need to create and look after such places?

CHAPTER SIXTEEN
A Charity Just for Bumblebees

If honey bees become extinct, human society will follow in four years.
Albert Einstein

Although this quote is oft-repeated, it is almost certain that Einstein did not actually say this. There is no record of when or where he said it, and I don't think he was prone to making sweeping statements on subjects in which he had no expertise. It is also almost certainly incorrect. It would undoubtedly be a disaster for some crops, and would put even more pressure on the global food supply which is steadily being stretched ever further by the growing human population. If the word 'honey' was removed from this quote, it would be a little more plausible. In the UK, honeybees contribute at most one-third of all insect pollination, with much of the remainder provided by wild bees including bumblebees. If we were to lose all of our bees, then our diets would be much poorer, although most of us would survive. The major sources of carbohydrate that support the human population are cereals – rice, maize, wheat, barley, sorghum, millet and so on – and all of these are pollinated by the wind rather than by insects. On the other hand, almost every fruit or vegetable that is good to eat is pollinated by bees; imagine a diet without almonds, blueberries, raspberries, beans, apples, melons, cherries, cucumbers, pumpkins, and many more besides. Many vegetables such as potatoes or cabbage do not require pollination by bees to produce an edible crop, but nonetheless pollination is needed to produce the seed for next year. Even cows require pollination. You may think I'm getting a bit carried away here, but many fodder crops, such as clovers and alfalfa, require bees to pollinate them and so without bees we would have fewer livestock and hence less meat.

Bees aren't just important for the foodstuffs they provide to us. A myriad of other organisms depend on bees, including the wild plants that they pollinate, the animals that feed upon those plants, the worms and woodlice that help to decompose those plants, the bacteria and fungi in the soil around their roots, and so on. All of our natural ecosystems would be radically altered and much poorer without insect pollinators, and in the UK, the predominant pollinators are bumblebees.

I began studying bumblebees not because they are important pollinators but because they are fascinating, because they behave in interesting and mysterious ways, and because they are rather lovable. But as I became more familiar with what was known about them, it was made clear that they were in urgent need of help. I read Sladen's book *The Humble-bee* from 1912, and Alford's *Bumblebees* from 1975, which describe the biology of the UK's twenty-five bumblebee species,* and it was soon apparent that many of the species that are described with familiarity in these books were now either extinct or had become exceedingly rare. The apple bumblebee,† Cullem's bumblebee and the

* Twenty-seven species have been described from the UK, but the cryptic bumblebee was not discovered here until very recently, and the tree bumblebee did not arrive until 2001.

† In truth the apple bumblebee may never have been resident in the UK. It is known only from four specimens, captured by one Frederick Smith and his son Edward on the dunes near Deal in Kent in about 1865. Smith was a highly experienced entomologist who worked at the British Museum, so the record is likely to be genuine, but none have been recorded since.

short-haired bumblebee had all gone extinct in Britain, and a further six species are on the UK 'Biodiversity Action Plan' – often abbreviated to UK BAP – in recognition of their endangered status. Some of these are perilously close to extinction; the great yellow bumblebee, once found throughout the UK, is now found only in the far north and west of Scotland, while the shrill carder bumblebee, formerly widespread in the south of England and Wales, is now known from just seven populations. So it was that my interest shifted from focusing on understanding the foraging behaviour of bumblebees, to working on the ecology of the rare species, with a particular focus on understanding why they had declined and what we might do about it.

The first question that sprang to mind was why had some species gone extinct while others seemed to be surviving reasonably well? What was different about the great yellow bumblebee, or the short-haired bumblebee, compared to the buff-tailed or common carder? A common cause of rarity and susceptibility to environmental changes is specialism. Were the rare species more specialised, perhaps with regard to where they collected their food, compared to the common ones? Together with my students I spent a number of years studying the flowers used by different bumblebee species, both common and rare, across the UK. We spent a lot of time on Salisbury Plain because it was one of the few places where we could observe some of the rare species. We went to the Somerset Levels, to the marshes of north Kent, the sand dunes of south Wales, and the Western Isles of Scotland to find out what the rare bees that survived in these places were feeding upon.

To cut a long story short, the answer was clover. Most of the rare species seemed to be very fond of clover, particularly red clover, and other wild legumes such as tufted vetch and bird's-foot trefoil, probably because these plants provide pollen that is unusually rich in protein. Most legumes are meadow plants, the sorts of species that I am slowly encouraging in my field in France. Many also have deep flowers, matched by the long tongues that most rare bumblebee species possess. Three of the four longest-tongued species in the UK are extinct or nearly so: the great yellow, short-haired and ruderal bumblebees. Of the very long-tongued bees, only the garden bumblebee remains widespread. The rare species also tended to emerge later from hibernation; while buff-tails, white-tails and common carders emerge in March and April, the rare species tend not to emerge until June. Overall, it seems that the rare species are flower-rich meadow specialists, with long tongues to feed on deep meadow flowers and a predilection for protein-rich pollen from legumes. Their late emergence corresponds neatly with the onset of flowering of the first meadow plants, usually in June. In contrast, the common species tend to be short-tongued and unfussy as to what they feed on. They emerge early from hibernation and feed upon the spring flowers often found in deciduous woodland, and on the many flowers found in gardens. By doing so they are able to get a head start on the late-sleeping species. Once we had established that our rare bees were specialists on meadow flowers, it was immediately obvious why they had declined so badly; almost all of our flower-rich meadows have been destroyed by modern farming methods.

During our gathering of data on the ecology of the UK's rare bumblebees, another aspect of their decline became apparent. Intensification of farming had driven these species from the wider countryside, confining them to pockets of flower-rich habitat which survived, such as Salisbury Plain or the machair of the Hebrides. But then these isolated populations had often become extinct, even though the habitat within them remained largely unchanged. Wicken Fen in Cambridgeshire provides a nice example. Wicken Fen is a well-managed National Nature Reserve which in 1920 supported fourteen species of bumblebee (excluding cuckoos). By 1978 only seven remained. There is no great mystery as to why this happened. If you stand on the edge of Wicken Fen looking outwards, all you can see is flat, intensive arable land stretching monotonously to the horizon. The Fen itself still has flowers, but not enough to support viable populations of anything but the most resilient bumblebees.

Bumblebees are at a distinct disadvantage compared to most other creatures when it comes to surviving on small nature reserves. Most bumblebees are sterile – they are workers – with only the queens and males producing offspring. The number of breeding females is equal to the number of nests, and each nest requires quite a bit of good habitat to support it. As an illustration, suppose a successful nest requires 1 hectare of high-quality habitat – I haven't yet managed to work out the exact figure, which is probably much larger, but this will do for the sake of argument. In contrast, 1 hectare of meadow could easily support 1,000 meadow brown butterflies, with 500 breeding females. You may be wondering why this is important. The answer is that population size is critical for long-term survival. You may read that there are only 720 mountain gorillas surviving in the wild, or sixty Javan rhinos, and that they are therefore critically endangered. As a ball-park figure, conservationists agree that a population of less than fifty is probably beyond reprieve, doomed no matter what is done to help it. There are two reasons for this: firstly, a small population might go extinct through simple bad luck – a predator could eat them all, poachers might shoot them, or a disease or flood could wipe them out. Secondly, small populations suffer from inbreeding, whereby after a few generations all individuals become genetically related. I mentioned in Chapter 9 that most of us carry dud copies of some genes, but that we have two copies of each and so long as we have one good copy we are usually fine. Mating with relatives – who are likely to have the same duff genes as ourselves – risks having offspring with two dud copies of the same gene, and that is a recipe for disaster.* This drop in offspring health is known as inbreeding depression, and can help to nudge an already vulnerable population towards extinction. Now most UK nature reserves are tiny, often only a few hectares, and so they are unlikely to be able to support more than a handful of nests of a rare bumblebee species, and a handful of nests does not constitute a viable population in the long term.

The pattern of extinction of the short-haired bumblebee illustrates this point. They were once widespread in southern England, but rapidly disappeared from most areas in the 1950s and 1960s as their habitat was swept away. By the 1970s only a half-dozen small populations remained, and one by one they went extinct. Dungeness National Nature Reserve was their last stronghold, but despite protection of the habitat at Dungeness, this last population faded away. It seems likely that such small, isolated populations were not big enough to be viable. If this is so, then are the remaining populations of the shrill carder and great yellow also heading steadily towards extinction, and if so, what can we do about it? It seems to me that we really need to get a handle on how small these populations are, and how much habitat is needed to support a viable population in the long term. It would also be really useful to know how far bees can travel between populations, for movement between populations helps to counter the negative effects of inbreeding by bringing in fresh blood at regular intervals.

* Charles Darwin experienced this phenomenon first-hand. The Darwin and Wedgwood families intermarried repeatedly over four generations, with Darwin marrying his cousin Emma. They had ten children, three of whom died early in life and three more of whom never had children of their own. It is likely that the taboos against marriage between relatives, which are widespread in human societies, came about to prevent such instances.

To try to get answers to these questions, I started a programme of genetic studies of populations of rare and common bumblebees which continues to this day. We focused initially on two BAP species, the moss carder and shrill carder, and one scarce but widespread species, the heath bumblebee. Funded by the Leverhulme Trust, Ben Darvill headed north to Scotland and the Hebrides to collect samples whilst a new student named Jon Ellis travelled the length and breadth of England and Wales. Their basic aim was to sample DNA from bees and use genetic markers to identify how many nests were in each surviving population, whether populations were suffering from inbreeding, and over what distances bees moved between populations. The Hebrides served as a really neat island system in which it was very easy to quantify how isolated each population was, since bees don't live in the sea. I won't bore you with all the details, but three years later we had some fascinating insights into the population biology of these species. Most worryingly, estimates of the surviving population sizes were very low. Jon sampled all the seven known populations of shrill carder and estimated the number of nests in each population to vary from twenty-two to twenty-nine. There were no doubt some nests that he failed to detect, but nonetheless this suggested that these populations were teetering on the edge of extinction. Moreover, the rare species had very low genetic diversity, a sign of inbreeding, and they seemed to have poor dispersal abilities. The surviving populations of shrill carder had little or no gene flow between them, suggesting that they were too far apart for any bees to fly from one to another. In Scotland, Ben's studies showed that moss carders also had low mobility, rarely flying as far as 5 miles between islands, while in contrast the more widespread heath bumblebee seemed to be able to traverse 20 or more miles of open sea on a regular basis. It seems likely that one factor that makes a species susceptible to decline is a poor dispersal ability, for this renders populations more likely to become isolated from one another.

All of this work is ongoing; we continue to gather information on the ecology of bumblebees, all of which helps us to understand what needs to be done to help support strong, healthy populations in the long term. It is clear that we urgently need to create more suitable habitat in and around surviving populations of rare species such as the shrill carder, so that these populations can expand rather than drift to extinction. Ideally we need to try to link these populations by providing habitat stepping-stones. We know what habitats are needed, what flowers they should contain, and we have a pretty good idea how to create them. However, there is a problem. Knowing what to do is all very well, but knowledge does not in itself conserve any bees. Most scientific papers are read only by other scientists, and not by the politicians, farmers, gardeners or nature reserve wardens who might be able to do something with the knowledge that researchers unearth. After a long day in the field, a farmer does not come home, have his dinner, then put his feet up in front of the fire to read the latest edition of the *Journal of Animal Ecology*. It is hard to imagine him calling out to his wife, 'Arrr, my luvver, avee zeen this article on dumbledores? Oyed better be zowin zum clover tomorrow.'*

Scientific articles are written to be understood by other scientists, and even then they often fail, for there are many papers that I struggle to make sense of. New methods of genetic analysis and statistics are constantly invented so it is very hard to keep up, and in any case over a million scientific articles are published every year, on average more than 150 on bumblebees alone. Even if a farmer or policy-maker had the time and background to make

* I must immediately apologise for this shockingly stereotypical portrayal of a farmer. Of course they are not all male or from Somerset, except in my head. Dumbledore, you may have gathered, is the old English name for bumblebee, possibly originating in Somerset or Sussex.

sense of these papers, he would be hard pushed to get hold of more than a small fraction of them, since most journals charge very substantial fees for access to their publications – even though the articles are written, reviewed and edited for free by scientists.

In short, one might find out everything there is to know about bumblebees, and publish it all for others to read in scientific journals, but only a handful of other scientists would read it and it would not result in there being one more bumblebee in the world. I found this enormously frustrating, and started to wonder what the point was. How could I actually do something practical to help? How could I persuade farmers to grow more flowers, or convince policy-makers of the need for agri-environmental policies that support bees and other wildlife, or convince staff in local councils to mow verges less frequently?

One day in 2005 it occurred to me that the solution might be to start a membership-based charity devoted to bumblebee conservation. After all, birds have the RSPB, butterflies have Butterfly Conservation – why shouldn't bumblebees have their own charity? I mooted the idea to my research group at a lab meeting in my office; they were a little bemused, but generally seemed to think it might be worth a try. I didn't have a clue what it could involve, but started to look into how one sets up a new charity. At the time I was still working at Southampton University, and I wasn't sure that the senior management would be at all keen on the idea. Academics are expected to bring in lots of grants and write heaps of scientific papers – these are the factors that bring funds and kudos to a university – so why would they want to pay my salary while I spent much of my time trying to save bumblebees? As luck would have it, I was already thinking of moving from Southampton after an eleven-year stint, and in June I was offered a professorship at Stirling University. At the time I was doing a lot of work on Hebridean bumblebees, so a move northwards seemed like a good idea. It was also clear that senior staff at Stirling were more imaginative and could see possible benefits of this new charity, if it took off, for the reputation of the university.

I moved up with my family in February 2006, initially renting a house in the village of Menstrie, a short cycle ride from the university. Cycling in on my first day was a strange feeling; the landscape, with towering hills all around, could not have been more different from Southampton. My office had a view over a loch with mountains beyond. It was also a shock to suddenly be on my own. At Southampton I had built up a team of students and postdocs, but they had opted to stay in the south, so I felt rather lonely in my office on that first day.

Shortly after I arrived I was contacted by a girl from the United States who wanted to do a PhD on bumblebees with me, and who had a wad of cash in the bank from a recent divorce and so was willing to fund herself. Eager to build up my group again, I took her on. Named Jennifer (known as Jenn) Harrison-Cripps, she lived up to two American stereotypes in that she had a very loud voice and an astonishing capacity for food, particularly chocolate-coated ginger biscuits. She also had the largest dog I have ever seen, a St Bernard that was slowly eating her house.

In the meantime, I had been preparing the groundwork for launching the charity. I couldn't think of a better title than the Bumblebee Conservation Trust, which had to be abbreviated to BBCT since BCT was taken by the Bat Conservation Trust. I registered BBCT as a company, and then applied for charitable status. I also opened a bank account. All of this was fairly straightforward. Rather speculatively, I filled in and posted off a very long application form for a grant of £50,000 from the Heritage Lottery Fund to employ a first member of staff. I checked out how other, similar, membership-based organisations worked, and realised that new members were usually offered a range of goodies to tempt them in. By dint of some blatant copying of what other organisations provided,

I came up with a metal pin badge of a bilberry bumblebee – designed using my children's felt-tip pens – a car window sticker, a pack of wild-flower seeds, and a bumblebee identification poster. For the latter, I persuaded an artist named Tony Hopkins, who had painted the lovely bumblebee pictures for the Naturalists' Handbook *Bumblebees*,* to let me use his pictures. I designed a logo, based on a bilberry bumblebee – the bee with the large red bottom – so we could print headed paper. I also cobbled together a newsletter, for which I came up with the title of *Buzzword* – cheesy, but I still like it.

I had no idea how many newsletters, pin badges and so on we would need, since I had no idea how many members we might get. It turned out that there were huge economies of scale; 500 pin badges were roughly twice the cost of 100.† Of course the charity had no money whatsoever at this point, so I had to pay for everything out of my own pocket, gambling that I would be able to reclaim it. I decided to go for 500 of everything, and within a few days all the goodies had been delivered in a selection of large cardboard boxes, which I stacked in my office. We were ready to go, but go where?

I wasn't sure what to do next. How were we going to get any members? The obvious option was to advertise, but advertising in any national media was prohibitively expensive since I would have had to pay for it myself. A press release seemed like the best way forward. The university has a press office, so I drafted something about the launch of a new charity to save our beleaguered bumblebees, and they sent it out. I had no idea what to expect. The following

* This book, by Oliver Prys-Jones and Sarah Corbet, is a great little introduction to British bumblebees, with a good identification guide, distribution maps and lots of drawings and pictures.

† This reminds me of my favourite adage, from an old home winemaking recipe book: 'Five gallons are as easy to make as one, and last nearly twice as long.'

day, Thursday 25 May 2006, I had a couple of calls from journalists on local papers who seemed interested, and finally, late in the afternoon, a call from Mike McCarthy at the *Independent*. I'd not heard of him at the time, but he is their environment editor. We chatted for a while, and I became quite excited as he was clearly planning to run a detailed piece. National newspaper coverage was exactly what we needed. Mike then asked me to hold for a few moments. I sat in my office at Stirling gazing at the view for quite some time before he came back on the line and dropped a bombshell. He said, 'What would you say to the entire front page?' I didn't say anything immediately as I was too busy falling off my chair in surprise. Mike was essentially offering for the *Independent* to sponsor a very prominent launch of BBCT. He wanted pictures and lots of information, and he wanted it immediately. This, as I have since discovered, is always the way with journalists, but I was hardly going to complain.

It turned into a long evening, but it was all worth it for the next day the *Independent* devoted not just the front page but the entire first three pages to bumblebees and the launch of BBCT. My phone didn't stop ringing for days, with other journalists wanting to do follow-up pieces and members of the public wishing to join the trust. I'd set standard membership at £16 per year, slightly undercutting other similar organisations, and within a day or two cheques started to flood in. It was at this point that the inadequacy of my preparation hit home. The trust had no staff, no office and no membership database. I had what was supposed to be a full-time job as an academic, and could only moonlight for the trust for a few hours a day. I had not anticipated how long it took to type names and addresses into a spreadsheet, to put together and send out membership packs, and to deal with the myriad of telephone enquiries, emails and letters that flooded in. To exacerbate the problem, I had to go to Portugal for eleven days in early June to teach on our undergraduate biology field course in a study centre with only the most rudimentary phone and Internet connections.

Jenn offered to step in, which was a massive relief. I left her in charge and headed off to the sun, asking her to send me email updates as to how many membership applications came in. Three days later I received an email from Jenn with a very large attachment. This wasn't very helpful as it took about an hour to download, but when it finally did it was a picture of her holding a vast armful of envelopes. Hundreds of membership applications were flooding in.

By the time I got back it was total chaos. It turned out that organisation was not Jenn's strong suit, although she had done her level best. My lab was a sea of opened envelopes, cheques and letters that were spilling off shelves, falling behind cupboards and spewing across the floor. Jenn had sent off membership packs to some, but she had lost track of which ones, so some ended up getting two packs, and there may well be others whose cheques were lost and who never received a thing. Of course it wasn't her fault; I should never have left her on her own to deal with all this, with no help and no systems in place. It took weeks to sort out, but thankfully the flood of letters slowed, and by the end of June we had 500 members and several thousand pounds in the bank. Thankfully our new members were mostly very tolerant of our ineptness.

During late summer of 2006 I managed to tempt Ben Darvill up to Scotland with the offer of a three-year postdoctoral position, and he became heavily involved with the trust, which took some of the pressure off. Then we had a very welcome piece of news: I received a letter from the Lottery saying that they were awarding us the £50,000. We advertised for the trust's first staff member, and were snowed under with applications. Membership was still growing slowly, bringing in a steady flow of money, and the applicants were so good that we decided to take a gamble and employ two rather than the one we had intended. As fate would have it, we ended up with Bridget England as our Scottish conservation officer, and the rather more appropriately named Lucy Southern as our England and Wales officer, albeit based in Scotland. The university kindly offered the trust an office for free, just along the corridor from myself, and we were up and running.

Even with two full-time staff, the first couple of years were hectic. None of us really knew what we were doing. We had to work out a finance system, produce annual financial reports, sort out tax, VAT, payrolls, pensions, deal with the growing membership and the many enquiries this generated, as well as try to deliver the conservation aims of the trust. Both Lucy and Bridget were trying to each do half a dozen specialist jobs, few of which they had any experience in. Ben and I tried to advise and help as much as possible, but we had no more experience in many of these areas than they did, and in any case we were fairly well occupied with our own academic posts. It is a marvel that the trust didn't implode, but somehow Bridget and Lucy managed to avoid any major disasters.

I'll spare you a blow-by-blow account of the intervening years. My role in the trust has diminished as it took flight and became self-sustaining. The trust now has eleven staff, and was recently awarded a new, much larger Lottery grant. Much more importantly, the trust has achieved far more than I could have ever dreamed of. It has 8,000 members all over the UK and a scattering overseas, and has been involved in creating in the region of 2,000 hectares of flower-rich habitat at sites from Orkney to Kent and Caithness to Pembrokeshire, most of it in our target areas where endangered bumblebee species occur. The trust hasn't tried to buy up land to manage, which would be inordinately expensive and would require staff and machinery that at the moment we simply don't have. In any case, the area of land needed to support a viable population of a rare bumblebee is huge – probably several square kilometres.

Instead, the trust has focused on working with existing landowners and managers, raising awareness of the presence of rare species in particular areas, and encouraging creation of new patches of flower-rich habitat wherever possible. Much of the 2,000 hectares of habitat has been created through encouraging and helping farmers to enter agri-environment schemes, whereby they can apply for government funding for creation and management of flower-rich grassland.

Most farmers have no idea that they have a rare bumblebee living nearby. For the last three years, the trust's current Scottish conservation officer, Bob Dawson, has focused his efforts on remedying this situation in areas where the great yellow bumblebee survives. This insect is neither as big nor as yellow as one might expect from the name, but it is nonetheless quite lovely, and was once found throughout Britain. In the last sixty years it has disappeared from about 95 per cent of its UK range – there are none left in England or Wales, and it is now found only on some Hebridean islands, Orkney and on the far north coast of Caithness and Sutherland. Bob has spent much of his summers for the last three years visiting these areas and talking to crofters and farmers. Most of them had not heard of a great yellow bumblebee until Bob arrived on their doorstep, but they were delighted to find that their land might be supporting such a rare creature, and more often than not they were willing to help by making the land better for bees. Many have since developed a sense of ownership of the great yellow; they are proud to be a part of conserving its future. Pippa Rayner, Lucy Southern's successor as English Conservation Officer has had similar experiences working with farmers in South Wales and Wiltshire to conserve the shrill carder. As the trust's capacity grows, our long-term aim is to work outwards from these remote sites scattered across the UK, hopefully ensuring the survival of the great yellow and the shrill carder and eventually encompassing all of the areas where rare bumblebee species survive.

On top of creating habitat, the trust has distributed more than 20,000 packs of wild-flower seed, and 4,000 copies of a little booklet I wrote on gardening for bumblebees. Gardens cover nearly 1 million hectares in the UK, so if we can make even a small fraction of them more bumblebee-friendly then we can have a major impact. Bumblebees seem to be doing pretty well in urban areas compared to the countryside, probably because there tend to be more flowers and nesting opportunities in gardens than in farmland. However, there is plenty of room for improvement; some gardens are terrible for bumblebees or any other wildlife. Low-maintenance decking, gravel, tarmac and concrete cover many, while carefully manicured lawns are also pretty much useless. Modern bedding plants have been intensively selected for size and colour, and in so doing they have lost their nectar, or become grossly misshapen or oversized so that it is impossible for bees to get to the rewards. As an example, the small, delicate wild pansy is popular with bumblebees, while the huge showy blooms of cultivated pansies are ignored. So-called 'F1 hybrids' are often sterile, having no pollen. 'Double' varieties have extra petals, which prevent bees from getting into the flower. For these reasons most of the busy lizzies, lobelias, petunias, begonias and so on that are sold as plug plants in the spring to provide an instant splash of colour are more or less useless to bees or butterflies; they have lost their original function, which was to attract pollinators.

Sad though all this is, it means that there is an opportunity to change gardens for the better. There are many flowers which are easy to grow and that are also great for bees. Many of the trust's members are keen gardeners, and vie with one another to attract the most bumblebees to their garden. In general, old-fashioned cottage garden perennials are the ones to go for, particularly garden herbs – lupins, hollyhocks, scabious, lavender, chives, sage, thyme and rosemary and so on. Most are easy to grow and low maintenance, so they are well suited to busy modern lifestyles – gardening for wildlife is easy. They are also beautiful, although perhaps not as showy as annual bedding plants. A wildlife-friendly garden does not have to be a chaotic mass of nettles and brambles.

Many gardeners do not realise the deficiencies of the bedding plants that they grow, and are happy to modify their habits once the benefits of more traditional plants are explained – it is just a matter of raising awareness. To try to get the message across, the trust is working with seed companies and garden centres to produce ranges of bee-friendly plants and has had stands at major gardening shows such as Hampton Court Flower Show. I bought a professionally made bumblebee outfit for such events, which is great at drawing attention to the bumblebee stand, but incredibly hot to wear for long. Younger children either love it or are terrified; on one occasion a small girl burst into tears when she saw me in the suit, so I pulled off the oversized head to reassure her that it was just a person inside. Unfortunately this made it worse, either because my face was more scary than that of the bee or because she thought that someone was stuck inside a giant bee, perhaps having been eaten.

Children are, of course, the gardeners, farmers and politicians of the future, so if we can encourage them to appreciate the importance of wildlife from an early age – rather than making them petrified of huge bees – perhaps we can influence the direction of society in the long term. To this end, we have developed an education pack aimed at primary schools in Scotland, which has gone out to over 200 schools and involved over 10,000 children in bumblebee-related activities.

On top of these activities, the trust has put on bumblebee identification walks, given talks in village halls, and organised 'farm days' for farmers to learn more about bumblebees. Many of these are run by trust staff but some of our more enthusiastic members have put on their own events. The trust staff have met with MPs and government ministers, and even been invited to Number 10. There have been at least two bumblebee-themed weddings, with all the guests wearing bumblebee pin badges and the cake decorated with marzipan bumblebees. Somehow we have discovered and tapped into a groundswell of affection for bumblebees which continues to grow.

For me, one of the most exciting new developments for the trust is the development of a 'citizen science' scheme to monitor bumblebee populations over time. Called 'Beewalks', this scheme is modelled on a similar and enormously successful butterfly monitoring scheme that has been running for nearly forty years. I'm slightly embarrassed to admit that we do not yet know which UK bumblebees are still declining, and which are not. Range declines, such as that of the great yellow, are easily recorded. But for widespread species such as the common carder, or the buff-tailed bumblebee, I cannot say whether they are less abundant now than they were ten or a hundred years ago. Logic suggests that they probably are, but there are no numbers to analyse and compare. Crucially, we do not know what the current trajectory is. This is vital information if we are to prioritise conservation efforts on the species that most need them, and if we are to measure the success of our conservation efforts. It thus seemed to me that we desperately needed to start counting bees as soon as possible, and so we launched Beewalks in 2010, advertising it to trust members, and 125 volunteers quickly signed up. They are required to walk a fixed route once per month during the spring and summer, counting how many of each bumblebee species they see, and then send in the data to Leanne Casey, my PhD student, who is coordinating the scheme. The aim is for Beewalks to continue indefinitely, and for the number of volunteers to grow, so that in the fullness of time we can build up a picture of the changing numbers of all of our bumblebee species across the UK.

After a faltering start, I am sure that the trust will continue to grow. I have no idea how large the membership base may one day become, and I cannot be certain that the trust will succeed in preventing further extinctions of bumblebees. There is much more to do. Perhaps the trust should one day be looking to spread its wings and start working overseas, for there are plenty of bumblebees elsewhere that need help, and in many countries there is precious little awareness of their plight.

Two things I do know. Firstly, the trust has raised awareness of bumblebees and their decline. Due to the many articles in the media that the trust has managed to generate since our launch in the *Independent*, there are now rather few people in the UK who are not at least dimly aware that bumblebees are in trouble and in need of our help. Secondly, there are also now many places in Britain where one can walk in a meadow full of flowers and happy bees, which would not be there were it not for the hard work of Lucy, Bridget, Bob and Pippa.

Conservation is not something that should be left to others. It is easy to get depressed and despondent at the impending extinction of the polar bear or the tiger, or at the horrific progress of deforestation in the tropics. Perhaps governments or scientists or organisations such as WWF can do something to help address these situations, but as an individual it is very hard to know where to start – it all seems so remote and dauntingly complex. In contrast, conserving bumblebees is something anyone can do. A single lavender bush on a patio or in a window box will attract and feed bumblebees, even in the heart of a city. Anyone with a garden can help enormously – plant some comfrey, viper's bugloss, foxgloves, chives, aquilegia and so on, and you will see the results almost immediately. If you are lucky enough to be a farmer, or a policy-maker, the warden of a nature reserve, or a planner in the local council, you can make a world of difference. This is not just about bumblebees, but about creating a future environment for our children to enjoy, where there are still flowers, bees, butterflies and birds, and healthy crops to eat.

CHAPTER SEVENTEEN
Return of the Queen

The presence of short-haired bumblebees in New Zealand had inspired a plan to bring them back to the UK. Mick and I had learned a fair bit about the sorts of plants that they needed by visiting them in New Zealand. In 2008 I arranged a meeting with the key players at the RSPB's headquarters in Bedfordshire (a beautiful and vast old stately home strangely known as the Lodge). Natural England's invertebrate specialist David Sheppard was there, along with Jane Sears from the RSPB, Mike Edwards and Paul Lee from an organisation known as Hymettus (which provides specialist advice on the conservation of bees, wasps and ants), and Brian Banks from Swift Ecology, a consultancy based in Kent. The final stronghold for the short-haired bumblebee had been Dungeness, with the very last British one seen there in 1988, so it made sense for this to be the first release site, should a reintroduction go ahead. It might seem odd that the RSPB was involved, but a substantial chunk of Dungeness is owned and managed by it as a nature reserve. It is perhaps not widely appreciated, but the RSPB makes great efforts to conserve species other than just birds, and had already been busy improving habitats for bumblebees on the reserve. Between us, we discussed what we knew about the short-haired bumblebee, and managed to convince ourselves that there was a realistic chance of success, so long as funding could be secured. It was clear that synchronising the bee with the UK climate was going to be a substantial obstacle, but we came up with a number of possible solutions to this. In the following weeks, David Sheppard pitched the idea to his bosses at Natural England and, to everyone's delight, they agreed to fund a three-year project, with sufficient money for a dedicated project officer.

We advertised the post and soon after appointed Nikki Gammans, a loquacious freckle-faced redhead from Essex. She had recently finished a PhD on ant biology, and had been involved in translocating rare ants back to sites from which they had died out, so she was well suited to the job; those of us who were on the interview panel had some reservations as to whether she would be able to relate to farmers since she had no farming background or experience, but they turned out to be ill-founded.

The first stage of the project had to be to create enough habitat for short-haired bumblebees to survive. After all, they had died out for a reason, and it would be very depressing and rather pointless to go to the expense of shipping them halfway round the world if they were just going to die out again.

Dungeness is a rather strange place. It has a peculiar, brooding atmosphere, no doubt in part due to the ugly concrete structure of the nuclear power plant that looms above it. Because of the extraordinary flatness of the landscape, the reactor and chimneys are always in view. Ecologically speaking, it is a very unusual habitat, one of the largest expanses of shingle in the world. It has been used for gravel extraction for many years, creating lots of shallow pools that are much loved by wading birds. One might imagine that shingle would be a fairly bleak and inhospitable habitat for bees, but in fact the shingle is swathed in an extraordinary diversity of flowers in spring and early summer. Because there are very few nutrients, legumes, which can fix their own nitrogen via root nodules, thrive, providing lots of the flowers that bumblebees adore. The shifting shingle provides a great habitat for viper's bugloss, a plant that also thrives on the stony overgrazed sheep pastures and pebble-strewn lake shores of New Zealand where we had found it to be a favourite with short-haired bumblebees. Why then had short-haired bumblebees died out here in the first place? Our best guess was that they had suffered due to changes on Romney Marsh, which encircles Dungeness on the landward side. The Marsh was once filled with flower-rich water meadows and hay meadows, which had been largely destroyed by intensive farming. The area of flower-rich shingle was probably not enough to support a viable short-haired bumblebee population, particularly since in dry years the flowers tend to die off on the shingle before bumblebee nests have completed their annual cycle. Thus the key to success with this reintroduction would be to replace some of this lost habitat on Romney Marsh and the surrounding area.

We could not be sure exactly how much good habitat would be needed, but there was no doubt that the more we could create the better the chances would be that the bees would survive. Thankfully, Brian Banks and local Natural England staff had been working for some years to encourage landowners in the area to improve habitats for bumblebees; although the short-haired bumblebee had died out, there were still other rare species in the area, such as the moss carder and brown-banded carder bumblebees. Together with the work done by the RSPB, there was already a fair bit of flower-rich habitat in the area – certainly more than there had been in 1988 when the short-haired bumblebee disappeared. Nikki set about creating more, working with local landowners, particularly farmers, to encourage them to put in pollen-and-nectar strips, or clover leys, or to sow wild-flower meadows. She organised 'farm days', when farmers could meet on a farm and see examples of flower-rich habitat and learn about short-haired bumblebees and the project. Just as the crofters in remote regions of Scotland quickly engaged with the idea of helping great yellows, so many farmers in Kent seemed genuinely excited at being involved in a project to bring back this extinct bee. A local wind-farm company also came on board, agreeing to sow a vast expanse of flowers under their turbines. In no time at all, patches of flowers were springing up all over Romney Marsh and around.

The next phase of the project was to investigate how we might get the bees back from New Zealand and into sync with the UK seasons. Queen bees are fairly easy to find and catch when they emerge from hibernation, since they spend several weeks flying about looking for somewhere to nest. Mick and I had seen short-haired bumblebee queens on our visit. But in New Zealand, these queens emerge from hibernation in December; if we caught them then and brought them back to the UK it would be midwinter and they would quickly freeze to death. Catching young newly mated queen bees at the end of the New Zealand summer (March) would be ideal, as these could be briefly hibernated, brought back and released in the UK three months later in June. The problem with this plan was that queens dive into hibernation underground almost as soon as they have finished mating, so they are seldom seen at the end of the summer. For a rare species such as the short-haired bumblebee in New Zealand, we were not optimistic that we would find enough by that route.

Of course the queens that were used for the original introduction to New Zealand were dug out of the ground while hibernating – if that were possible during the New Zealand autumn, then they could be brought back to the UK while still hibernating and woken up early in June. However, in New Zealand we had seen only small numbers of short-haired bumblebees scattered across a vast area of stony countryside. We had no idea where to dig for the hibernating queens, since there was no obvious equivalent of the ditch sides from which short-haired bumblebee queens were originally dug in Kent. Digging holes randomly to look for queens would be a back-breaking and utterly futile exercise. What then to do?

The ideal solution would be to catch nest-searching queens in the New Zealand spring (December) and persuade them to rear nests in captivity. If a number of nests could be reared then both new queens and males would be produced in March, and these could be mated in cages, the queens put into hibernation and then shipped back to the UK in refrigerated conditions for release in June. This would have been very easy for buff-tailed bumblebees, which breed readily in captivity. Unfortunately many other bumblebee species are extraordinarily hard to breed in captivity, and very little information was available as to how to breed short-haired bumblebees. Nikki did track down a Czech bumblebee enthusiast, Vladimír Ptacek, who had reared one or two short-haired bumblebee nests, and she visited him to find out the details. He had done it by placing young nests in large cages full of flowering clover so that the bees could collect their own food. This was all very well but meant that the rearing would have to be done in New Zealand, for we would not be able to provide stands of flowering clover in the British winter. While investigating the possibilities we stumbled across contact details for a Rosemary Reid who lived in Christchurch on South Island and who bred bumblebees semi-professionally, selling the nests to farmers. She had apparently bred short-haired bumblebees in the past, and was willing to rear our bees for us, for a price. We agreed that Nikki would go to New Zealand to catch the queens in December 2009, and supply them to Rosemary to set up the captive breeding programme.

With the help of local New Zealand bumblebee expert Barry Donovan, Nikki had little trouble collecting queens, which she stored in hair curlers* in the fridge in her camper van and transported back to Rosemary's house in Christchurch. Rosemary gave each queen a supply of nectar to drink and a ball of pollen mixed with nectar in which to lay her eggs. For some she added workers of garden bumblebees, in the hope that they would help the queen rear her brood. Via Nikki, we received regular reports as to the progress of the queens. Some did not take to captivity and soon died. Others seemed to settle down and lay eggs, but then died unexpectedly. Still others successfully reared some offspring, but their nests grew painfully slowly. Part of the problem may have been that Rosemary didn't have fresh clover pollen, which we think is their favourite, to feed them at the start. Whatever the reasons, the number of queens steadily dwindled until just a handful remained. A few finally produced males and just five new queens, which readily mated. These five queens were put into hibernation prior to being shipped back to the UK. Five wasn't likely to be enough to establish a new population, but at least we could gain experience in bringing them back, or so we thought. In the meantime their dead nest mates were sent to Mark Brown, a bumblebee disease expert at Royal Holloway in London, to make sure that they didn't have any unpleasant diseases that we would not want to accidentally bring into the UK. Sadly, a few weeks later Rosemary emailed to say that the five queens had died of unknown

* Strangely, the old-fashioned hair curlers, plastic open-ended cylinders with lots of small holes and protruding nobbles on the sides, are perfect for temporary storage of bumblebee queens. Each hair curler is stoppered at both ends with the queen inside, and then they can be packed together with dental wadding between them, the latter soaked in sugar solution. So long as they are kept cool, queens can survive for a week or so like this, occasionally poking their tongues into the dental wadding for a refreshing drink.

causes in hibernation. There would be no reintroduction of short-haired bumblebees in 2010.

Nikki came home from New Zealand and spent her summer encouraging the creation of more flowery habitat in Kent. In December, she went back to New Zealand to try once more. Breeding the bees in captivity had not been a roaring success, and Rosemary was asking for considerably more money this year to repeat a process that had been decidedly fruitless. We couldn't afford to pay what she was asking, and we were not convinced that she would do any better second time around, so we decided on a different tack. Bumblebee nest boxes are notoriously ineffective in the UK, but in New Zealand they seem to work quite well, and there are old records of them being used by short-haired bumblebees as well as the more common bumblebees found in New Zealand. If the bees could be persuaded to nest in an artificial box outside, then the nests would look after themselves and could be collected in just as they started to produce new queens and males.

With the help of Barry, Nikki set out nest boxes at the sites where she had seen most short-haired bumblebees the previous year. She monitored them every few days, but disappointingly she saw no short-haired queens anywhere near them. After a week or two she started to get a little desperate, and experimented with catching queens and confining them in the boxes with food. This has sometimes been found to work with other species; once they have been trapped in a nest box for a few days they grow used to it so that when the door is opened they do not simply fly away, but adopt the nest as their own. Unfortunately Nikki's bees had not taken to their boxes, and promptly disappeared when she opened the door. Within a few weeks the time when queens start nesting was over, and we had nothing to show for it. Nikki tried searching for wild nests but it was like looking for a needle in a haystack. Eventually, rather despondent, she returned home.

In the meantime, back in my lab in Stirling, Gillian Lye had been studying the DNA of short-haired bumblebees. I had previously brought back toe samples from my visit to New Zealand with Mick. Gillian had also visited the Hope Entomology Collection in Oxford where she had been allowed to take toe clips from short-haired bumblebee specimens collected before their extinction in the UK. Finally, she had asked a Swedish scientist, Bjorn Cederberg, to send her samples from the only known strong European population of this species, in southern Sweden. Gillian used genetic markers to study the amount of genetic diversity in each population – a measure of their genetic health – and also to compare how similar the three populations were to one another. Her results were somewhat alarming. The New Zealand bees were decidedly weird. They had astonishingly little genetic variation, and were very different from the UK museum specimens. With the help of Olivier Lepais, a French expert in genetic analyses who was briefly based at Stirling at the time, she was able to estimate the probable number of short-haired bumblebees that were introduced from Kent in 1885. You may recall that ninety-seven queen bees survived the journey to New Zealand and flew away when released at the Christchurch Botanic Gardens. No record was kept as to which species were included, but it seems likely that the majority would have been of the more common species such as the buff-tail. Although four species of bumblebee became established in New Zealand, it is very probable that the sample also included other common UK bumblebees such as the white-tail. It would be a fair guess that there were rather few short-haired bumblebees since this was always a relatively uncommon species, but nonetheless Gillian's results came as a bit of a shock. Her data suggested that the entire New Zealand population of short-haired bumblebees was descended from just two queens.

This is an extreme example of what is known as a genetic bottleneck. If a population crashes to very low numbers – or in this case is founded by very small numbers of individuals – then it loses most of the genetic variation present in the original population. This reduces the potential of the population to adapt, and also leads to very strong inbreeding, since all the individuals are closely related to one another. If the population was started by two queens then within just one generation all bees would be brothers and sisters or at best cousins of one another. As we have seen, inbreeding leads to the expression of rare, recessive and harmful genes, which can result in deformities and generally low survival of individuals. Such populations would normally be expected to die out swiftly, but just occasionally, as here, they do not.* In New Zealand, there are lots of flowers, and rather fewer competitors than in England. Also, most of the diseases of bumblebees were left behind, so life has probably been fairly easy for bumblebees in New Zealand. It would seem that even bees of low genetic health were able to survive there. But would these bees be able to cope back in England, when faced with more competition and when exposed to diseases that they had not encountered for well over 100 generations?

* Hedgehogs in New Zealand provide a neat example. They were introduced before bumblebees in small numbers, and the resulting inbreeding has led to them having unappealingly malformed teeth. Nonetheless, in the absence of competitors they survive very well, and have made an enormous nuisance of themselves. Even with their sub-standard dentition they cheerfully and effectively consume the eggs of endangered birds such as the black stilt and black-fronted tern. One hedgehog was found to contain 283 weta legs; these are fearsome-looking mouse-sized crickets, found only in New Zealand, which are rapidly heading towards extinction thanks to our prickly friends and other introduced enemies.

Further doubts were raised when we examined the patterns of relatedness between the three populations that Gillian had examined. Aside from being inbred, the New Zealand bees were very different from those from the UK. That initial bottleneck and then 126 years of isolation had wrought huge changes in their genetic make-up. Ironically, the Swedish bees were more similar to the original UK bees than were those from New Zealand, despite that fact that the New Zealand bees were direct descendants of the UK population.

Gillian's work caused us to rethink our plans. Should we continue with our efforts to bring short-haired bumblebees back from New Zealand, or switch to an introduction from Sweden? The genetic data suggested that the New Zealand stock was in pretty poor shape, and that the Swedish bees were actually closer to the original UK bees, but nonetheless there was resistance to the switch. There was a beautiful symmetry to the idea of bringing these bees back to the UK from the other side of the world after a 126-year absence. David Sheppard of Natural England was initially reluctant, correctly arguing that we would probably have not considered doing the reintroduction at all if not for the existence of the New Zealand population. Natural England wanted British bees or bust. However, after long debate, they came around to the new approach. It has significant advantages. The Swedish bees are not out of synchrony with our seasons, making the process much easier since we could simply catch spring queens in Sweden and ship them direct to Kent for release. The source population is relatively healthy in genetic terms, and similar to the original UK population. Sweden has the same range of bee diseases as the UK, so the bees would not be exposed to anything new on arrival; this also made it unlikely that we would accidentally import a disease strain that might harm native UK bees. Overall, it seemed much more likely that the reintroduction would succeed if we used Swedish bees.

By the time this decision was made it was too late to organise a release for 2011, but buoyed by the realistic prospect of a release in June 2012, Nikki returned to her spring and summer job of encouraging landowners to create habitat around Dungeness. This work went extraordinarily well. I'm not sure how, but by the end of the summer of 2011 Nikki and the project partners had helped to create over 500 hectares of new flower-rich habitat in south-east Kent. Some farmers had put whole fields into red clover leys, many had sown strips of wild flowers along field margins, and others had undertaken major meadow restoration projects. Brian Banks produced maps of the new habitat, which showed a rash of patches around Dungeness, westwards to Rye and north-west to the edge of the High Weald of Sussex. What is more, we gained strong evidence that this work was benefiting bumblebees. Nikki, Brian and a team of volunteers had been recording bumblebee numbers around Dungeness for the previous few years. Dungeness and Romney Marsh used to have one of the richest bumblebee faunas in Britain, but it wasn't only the short-haired bumblebee that died out in the area. Shrill carders, red-shanked carders and ruderal bumblebees had also vanished. In 2011, all three of these reappeared of their own accord; several ruderal bumblebees were found in the west around Rye harbour, a single shrill carder turned up at Dungeness, and a red-shanked carder was spotted to the north-east of Rye. What is more, two other endangered species which had clung on in the area in small numbers, the brown-banded carder and the moss carder, had both increased and extended their ranges in the region.

In mid-April 2012 Nikki travelled over to Sweden, armed with her butterfly net and a large box of hair curlers. We all knew that this was really too early for short-haired bumblebees, which don't usually emerge from hibernation until well into May, but Nikki was champing at the bit and didn't want to miss them if it turned out to be an unusual year and the queens emerged early. We had gathered all the necessary permissions to capture the bees in Sweden, but nonetheless a local conservationist somehow got the impression that we had not, and by the power of social media quickly mounted a campaign to prevent the capture going ahead. There was a brief storm of media interest in the Swedish newspapers, in which we were described as 'typical British imperialists', amongst many other things, but thankfully this quickly died away when it was discovered that we did have both formal permission and the support of local bumblebee experts.

For reasons we do not fully understand, short-haired bumblebees seem to be thriving in southern Sweden, unaffected by the problems that have beset them elsewhere in Europe. Nikki had no trouble catching queens, most of which were feeding on white dead-nettle along the verges of country lanes. By 10 May, eighty-nine of them were safely housed in hair curlers in the camper van and were on their way back to the UK. Nikki handed them over to Mark Brown, who held the queens in quarantine for a fortnight while he carefully examined their faeces to determine if any of the bees were carrying infections. During this period a few of them died, with a parasitoid wasp named *Syntretus* bursting from their bodies. A few more were found to be infected with gut diseases and had to be humanely destroyed. On 28 May, Nikki picked up the remaining fifty-one healthy bees and drove them to Dungeness for their big day.

An excited crowd had gathered at the appointed spot, on the edge of a field full of clover. The respective press offices of Natural England, the RSPB and the Bumblebee Conservation Trust had done a great job and there were swarms of journalists and camera teams, as well as local conservationists, farmers and staff of the various organisations. Nikki had placed a few of the queens in clear plastic boxes, but kept them cool and in the dark until the last moment so that they did not become flustered and bash themselves about trying to escape. When the cameras were ready, and to a hush from the audience, Nikki revealed the first queen bee. Shutters clicked eagerly, and then the lid was opened. The queen buzzed her wings once or twice experimentally, warming her flight muscles while she tasted the air with her antennae. And after a moment or two, she flew away. For the first time in twenty-four years, a short-haired bumblebee was on the wing in England.

The story was much the same with the rest. A few paused to feed on the clover, but then they were off, heading away across the flat fields. And that was the last we saw of them. Of course this was exactly what we might have expected them to do, but nonetheless it seemed rather anticlimactic. Slowly, with nothing left to look at, the crowd of people dispersed and it was all over, at least as far as the media was concerned.

For the rest of the summer, Nikki and her team of volunteers searched for the queens and their offspring, but none were seen. The summer of 2012 was spectacularly wet and awful, which was unfortunate and may have led to the demise of our precious Swedish bees, or it may not. Queen bees are powerful fliers and we have no idea how far they might have gone, or what their fates were. They could have been more or less anywhere in south Kent or east Sussex, and looking for fifty-one bees across several hundred square miles of countryside is a thankless task. It would be easier if the short-haired bumblebee were a distinctive species, but they do look rather similar to several other native species so only an experienced person is likely to notice one, and there aren't many experienced bee spotters. In fact, many members of the public claimed to have seen one having read about the release in the papers, and some sent in photographs of what turned out to be other bumblebee species. None of the sightings could be confirmed.

As I write this last chapter, in a very wet January 2013, we have no idea if there are short-haired bumblebees alive in the UK. If there are, they will be the daughters of the queens that were released, tucked away somewhere underground awaiting the spring. If they become established somewhere in Kent then their numbers should increase, and before long we are bound to see them. Whatever happens, the funding from Natural England is ongoing for the moment and Nikki plans to release more queens from Sweden in 2013. Success is far from certain; this has never been done before. If all goes well, at some point we should see workers, and then we will know that a queen has successfully built a nest. Better still would be to see males and fresh young queens in late summer. The long-term plan is to develop further sites, starting with north Kent, with the eventual aim of establishing a network of linked populations in the south-east of England. To do this will necessitate creating massive amounts of new habitat, which will benefit other bumblebees, the wild flowers on which they depend, and many other wild creatures. Although these bees may never come back from New Zealand, the rather peculiar series of events that led to the existence of British short-haired bumblebees on the other side of the world has stimulated a project that has resulted in sweeping benefits for wildlife in one corner of England, and perhaps these benefits can spread. This rather nondescript bee is acting as a flagship for conservation efforts in the region, and bumblebees can do the same for conservation across the UK and beyond. Their direct relevance to man through crop pollination makes it very easy to explain the importance of conserving them, and from there it is but a small step to explaining that our survival and wellbeing is inextricably linked to that of all the wonderful diversity of life on earth. We need worms to create soil; flies and beetles and fungi to break down dung; ladybirds and hoverflies to eat greenfly; bees and butterflies to pollinate plants;

plants to provide food, oxygen, fuel and medicines and hold the soil together; and bacteria to help plants fix nitrogen and to help cows to digest grass. We have barely begun to understand the complexity of interactions between living creatures on earth, yet we often choose to squander the irreplaceable, to discard those things that both keep us alive and make life worth living. Perhaps if we learn to save a bee today we can save the world tomorrow?

The last word in ignorance is the man who says of an animal or plant, 'What good is it?' If the biota, in the course of aeons, has built something we like but do not understand, then who but a fool would discard seemingly useless parts? To keep every cog and wheel is the first precaution of intelligent tinkering.

Aldo Leopold

APPENDIX

Common and Latin Names of British Bumblebees

Some of my fellow scientists may be a little annoyed by my persistent use of common names for bumblebee species. This is frowned upon as it can lead to confusion: different common names are used in different countries, and sometimes multiple common names may be in use. On the other hand, a blizzard of Latin names can be rather off-putting. Below are the formal Latin names and English names of the British bumblebee species. The English name for the cryptic bumblebee I made up, but it seems a logical translation of the Latin, and appropriate enough for a bee that can be distinguished from other species only by examining its DNA.

Bombus barbutellus	Barbut's cuckoo bumblebee
Bombus bohemicus	Gypsy cuckoo bumblebee
Bombus campestris	Field cuckoo bumblebee
Bombus cryptarum	Cryptic bumblebee
Bombus cullumanus	Cullem's bumblebee
Bombus distinguendus	Great yellow bumblebee
Bombus hortorum	Garden bumblebee
Bombus humilis	Brown-banded carder bumblebee
Bombus hypnorum	Tree bumblebee
Bombus jonellus	Heath bumblebee
Bombus lapidarius	Red-tailed bumblebee
Bombus lucorum	White-tailed bumblebee

Bombus magnus	Northern white-tailed bumblebee
Bombus monticola	Bilberry bumblebee (blaeberry bumblebee in Scotland)
Bombus muscorum	Moss carder bumblebee
Bombus pascuorum	Common carder bumblebee
Bombus pomorum	Apple bumblebee
Bombus pratorum	Early bumblebee
Bombus ruderarius	Red-shanked carder bumblebee
Bombus ruderatus	Ruderal bumblebee or large garden bumblebee
Bombus rupestris	Hill cuckoo bumblebee
Bombus soroeensis	Broken-belted bumblebee
Bombus subterraneus	Short-haired bumblebee
Bombus sylvarum	Shrill carder bumblebee
Bombus sylvestris	Forest cuckoo bumblebee
Bombus terrestris	Buff-tailed bumblebee
Bombus vestalis	Southern cuckoo bumblebee or vestal cuckoo bumblebee

Acknowledgements

Particular thanks are due to my agent Patrick Walsh for his support and encouragement, and to Ellen Rotheray and Kirsty Park, my first and best reviewers. I'd also like to thank all of my research collaborators and my students for their ideas and enthusiasm, and to ask their forgiveness for any inaccuracies in my recall of events. Thanks are also due to my wife, Lara, and our three lovely boys, who somehow put up with my considerable eccentricities.

Index

About the Author

DAVE GOULSON studied biology at Oxford University and is now a professor of biological sciences at the University of Sussex. He founded the Bumblebee Conservation Trust in 2006, whose ground-breaking conservation work won him the Heritage Lottery Award for Best Environmental Project and the Social Innovator of the Year Award from the Biology and Biotechnology Research Council in 2010.